TELECOMMUNICATIONS RESEARCH RESOURCES

An Annotated Guide

TELECOMMUNICATIONS

A Series of Volumes Edited
by Christopher H. Sterling

TELECOMMUNICATIONS RESEARCH RESOURCES

An Annotated Guide

James K. Bracken
The Ohio State University

Christopher H. Sterling
The George Washington University

LEA **LAWRENCE ERLBAUM ASSOCIATES, PUBLISHERS**
1995   Mahwah, New Jersey   Hove, UK

Lawrence Erlbaum Associates, Inc., Publishers
10 Industrial Avenue
Mahwah, New Jersey 07430

Library of Congress Cataloging-in-Publication Data

Bracken, James K., 1952-
    Telecommunications research resources: an annotated guide / James
K. Bracken, Christopher H. Sterling.
        p.   cm.
    Includes index.
    ISBN 0-8058-1886-3 (alk. paper). -- ISBN 0-8058-1887-1 (pbk. :
alk. paper)
    1. Telecommunication--Research--Handbooks, manuals, etc.
2. Telecommunication--Research--Bibliography. 3. Bibliography-
-Bibliography--Telecommunication. I. Sterling, Christopher H.,
1943-   . II. Title.
TK5102.7.B73 1995
384'.072--dc20   95-9653
                                                              CIP

Printed in the United States of America

10 9 8 7 6 5 4 3 2 1

Contents

Introduction

As the telecommunications and information field expands and becomes more varied, so, too, do publications about these technologies and industries. This book is a first attempt to provide a general guide to that wealth of English-language publications, both books and periodicals, on all aspects of telecommunication. Though we focus primarily on the domestic American industry and policy, one chapter details the more important international publications. The book is conceived as a short-cut or road map for researchers of all types who need to find specific information.

What this book covers...

We define telecommunications as including all means of electrical and electronic communication *except* for electronic mass media. Within that admittedly broad definition, taking in wire and wireless telecommunications, this volume offers an indexed inventory of over a thousand resources on most aspects of American telecommunications history, technology, industry and economics, applications and impact, domestic and international policy.

While we make no claim to a complete census of all materials, the most important are here as a good indication of what is available. Included among the variety of sources discussed are organizations, databases, traditional books and reports, government entities, key periodicals, statistical reports, archives and museums and their major unpublished collections, and corporate and employment data.

This guide attempts to deal with telecommunication's extensive primary and secondary resources by striking a practical balance between, on the one hand, a sourcebook (or directory) for the major entities that comprise the American telecommunications infrastructure and, on the other, a comprehensive listing of *sources about* telecommunications. Our course has necessarily been one of selectivity. For example, we only identify telecommunication's major corporate and academic research organizations; trade and industry, professional, and public interest organizations and groups; and federal, state, multistate, and multinational regulatory agencies and other organizations. We believe that these and other entities give shape to the local, national, and international landscapes of U.S. telecommunications. Our annotations indicate that many of these companies, groups, and offices significantly add to telecommunication's literature. Meanwhile, we identify compilations of company or investor reports rather than references to the individual documents, guides to sources of statistics rather than the data themselves, directories of consultants rather than the names of specific consulting firms, etc. To provide specifics for these and other aspects of U.S. telecommunications would be to seriously date this volume even before its publication.

...and What it Does Not

Generally speaking, this book excludes resources devoted to:
- electronic media (broadcasting, cable, and related fields, which are covered in a companion volume)
- computers and information science (though it is often difficult to divide these from telecommunications and when in doubt or an item overlaps both, we include it here)
- highly technical engineering works

1

• languages other than English (with occasional exceptions when nothing in English is available or the item is unique)
• foreign telecommunications systems
• international telecommunications other than resources that illustrate the American role
• specific Congressional hearings on telecommunications (but see 0738 and 0739).

In the cases of electronic media, computers and information science, and high technology, at least adequate bibliographic resources already exist and to include such topics would make for an unwieldy volume. At the same time, these topics cannot be neatly distinguished or separated from telecommunications as we try to define it here: they are integral to telecommunication's context. Thus, throughout this guide we include entries for resources in the several fields and disciplines that represent this interdisciplinary—indeed, multidisciplinary—overlap with telecommunications.

The non-English language literature dealing with U.S. telecommunications and foreign telecommunications systems are here regarded as separate and different problems. Both literatures are certainly more voluminous than many American scholars might realize, largely because they are substantially unknown in the United States; that is, more accurately, non-English language works on U.S. and foreign telecommunications systems are mostly unavailable in U.S. research libraries and unrecorded in many of the field's indexes, bibliographies, and research guides, including COMMUNI-CATIONS ABSTRACTS (0016), Blum and Wilhoit's MASS MEDIA BIBLIOGRAPHY (0001), Sova and Sova's COMMUNICATION SERIALS (0005), and Sterling's TELECOMMUNICATIONS, ELECTRONIC MEDIA, AND GLOBAL COMMUNICATIONS (0006), as well as Eleanor S. Block and James K. Bracken's COMMUNICATION AND THE MASS MEDIA: A GUIDE TO THE REFERENCE LITERATURE (Englewood, CO: Libraries Unlimited, 1991), Rebecca B. Rubin, Alan M. Rubin, and Linda J. Piele's COMMUNICATION RESEARCH: STRATEGIES AND RESOURCES, 2nd ed. (Belmont, CA: Wadsworth, 1990), Jean Ward and Kathleen Hansen's SEARCH STRATEGIES IN MASS COMMUNICATION, 2nd ed. (New York: Longman, 1993), among others. Among standard resources in the field, the exception to this is COMMUNICATION BOOKNOTES: RECENT TITLES IN TELECOMMUNICATION, INFORMATION, AND MEDIA (Annandale, VA: Sterling, 1969-date), a monthly review journal regularly featuring full bibliographic data on selected new publications from around the world, with an emphasis on Canadian and European imprints. Nonetheless, a quick assessment of entries in Chapter 7 for indexing, abstracting, and biblio-graphical resources like COMMUNICATION CONTENTS SISALLOT (0930), EXCERPTA INFOR-MATICA (0931), INDICATOR CONTENTS (0932), MASS COM PERIODICAL LITERATURE INDEX (0934), and NORDICOM: BIBLIOGRAPHY OF NORDIC MASS COMMUNICATION LITERATURE (0935), as well as acknowledgement of other major sources, like Wilbert Ubbens' massive JAHRESBIBLIOGRAHIE MASSENKOMMUNIKATION (Berlin: Wissenschaftsverlag Volker Speiss, 1982–date, annual), suggests the voluminousness of non-English language research in the field. That many of these resources have been published for over a decade and yet are mostly unavailable in U.S. research libraries, we suggest, indicates that foreign scholars are doing a far better job in bibliographically controlling telecommunication's international literature than U.S. scholars. We do not attempt to remedy this problem: that is the stuff of a larger compilation.

Entries, Annotations, and Arrangement

In general, entries for published works cite author(s) or editor(s), title and subtitle, place of publication, publisher, publication year(s), edition, and inclusive pagination. Entries for organizations and other entities give the organization's name, main address, and telephone voice and fax numbers. Annotations generally identify former titles and organization names. In particular, entries for reference works and journals also indicate availablity and coverage of electronic versions as well as such features as particu-larly relevant sections or subject headings. Entries for secondary works note the inclusion of useful figures, illustrations, bibliographies, indexes, etc. In general, entries for organizations indicate any useful journals or other publications as well as guides or finding aids for resources issued by or about the organizations. A notable exception to this relates to museums, archives, and libraries included in Chapter 2.B (see p. 17).

Given the number of resources included, annotations are indicative of an organization's or resource's focus or content rather than an exhaustive inventory of specific details. Additionally, comments are intended to place an organization or a work in the context of U.S telecommunication's infrastructure or in the context of U.S. telecommunication's research literature. Annotations aim at objectivity; on the other hand, given that inclusion is limited by selectivity, we admit that we feel that everything included here is at least important and in many cases essential.

As the detailed table of contents makes clear, this volume is arranged in topical order to facilitate finding sources on related subjects. Each chapter begins with the basic reference literature— book-length bibliographies, indexes and abstracting services, directories, yearbooks, dictionaries, and the like. Each chapter concludes with a brief survey of the more valuable secondary sources (chiefly books) useful for background or a broad introduction. These lists focus on titles published within the past decade, with a scattering of classic and still-useful older titles.

Entries appear but once, though we make use of extensive cross-reference by entry number.

We have provided appendixes that detail Dewey Decimal and Library of Congress classification systems for telecommunications and Library of Congress subject headings related to telecommunications.

Keeping Current

Any published bibliography, by its very nature, is dated before you hold a copy in your hands. Given the pace of publication of new and revised resources, in addition to many of the standard resources identified in Chapter 1 (especially) and in the topical listings for bibliographies and indexes throughout this guide, we strongly recommend COMMUNICATION BOOKNOTES: RECENT TITLES IN TELECOMMU-NICATION, INFORMATION, AND MEDIA, noted above, for full bibliographic data and desciptions of selected new publications in telecommunications from around the world.

Our Thanks.....

With thanks (now that it's over) to ... Susan Hill of the National Association of Broadcasters as a model of what a trade association reference librarian can and should be ... research librarians at GWU's Gelman Library ... librarians of The Ohio State University Libraries ... librarians of the Libraries of the University of Amsterdam, The Netherlands, especially Drs. J.J.M. De Haas and Josein Huizinga, for help with international resources and for their hospitality... Jennifer (Spot Color) for typesetting ... and, of course, Ellen, and Cory...

About The Editors

James K. Bracken is Associate Professor in the Ohio State University Libraries where he has been bibliographer for communication since 1988. He is an Associate of Ohio State's Center for Advanced Study in Telecommunications (CAST) and the co-author (with Eleanor S. Block) of COMMUNICATION AND THE MASS MEDIA: A GUIDE TO THE REFERENCE LITERATURE (1991) and author or co-author of other reference works and articles on subjects ranging from telecommunications to early printing history. He holds a Ph.D in English literature and M.L. (Master of Librarianship) from the University of South Carolina.

Christopher H. Sterling is Associate Dean for Graduate Affairs in Columbian College of Arts and Sciences and a professor of communication at George Washington University. He has served on the GWU faculty since 1982, directing its graduate telecommunications programs from 1984–1994. Author or co-author of a dozen books including both reference and text books in both electronic media and telecommunications, he continues to edit COMMUNICATION BOOKNOTES which he founded in 1969. He holds a Ph.D in communication from the University of Wisconsin, Madison.

General Reference

This first chapter identifies major resources for research on telecommunications in general. More importantly it includes selected standard general resources that place telecommunications in a larger context. Likewise, we have identified several electronic services recently made widely available in the Online Computer Library Center (OCLC) bibliographic utility. None of these can be regarded as specific telecommunications resources as each covers far more than telcommunications' literature, organizations, or personalities. In light of telecommunications' dynamically changing landscape, however, general resources like many of those in this chapter serve to measure telecommunications' inevitable integration and convergence in many aspects of daily life.

1-A. Bibliographies

0001　Blum, Eleanor, and Frances Goins Wilhoit. **MASS MEDIA BIBLIOGRAPHY: AN ANNOTATED GUIDE TO BOOKS AND JOURNALS FOR RESEARCH AND REFERENCE.** Urbana: University of Illinois Press, 1990 (3rd ed.), 344 pp. Standard bibliographic guide to studies on printed and electronic media. Includes about 300 general items related to telecommunications as well as selected major indexes and English-language journals. Subject index.

0002　**DIRECTORIES IN PRINT.** Detroit: Gale, 1989–date, annual. Formerly DIRECTORY OF DIRECTORIES (1980–1988). Online version GALE DATABASE OF PUBLICATIONS AND BROADCAST MEDIA available from Dialog: covers current edition, updated semiannually. Diskette version available from Gale. 11th ed. (1993) for 1994. Classified descriptions of professional, trade, and other listings for membership, personnel, or companies, buyer's guides to services, products, ratings and rankings lists, and other kinds of directories. Relevant entries in chapters 7 for telephone utilities, 15 for engineering and technology, and especially 12 for "Telecommunications and Computer Science," including "data processing and computer research, manufacturing, retailing, installations, and user groups; telecommunications systems and research." Title and keyword indexes; subject index cross references entries under "Cable television broadcasting industry," "Satellite communications," Telecommunications," "Telephone industry," and others.

0003　**GUIDE TO REFERENCE BOOKS.** Eugene P. Sheehy, editor. Chicago: American Library Association, 1986 (10th ed.), 1,560 pp. See next entry.

0004　_____ **COVERING MATERIALS FROM 1985–1990: SUPPLEMENT TO THE TENTH EDITION.** Robert Balay, editor. Chicago: American Library Association, 1992, 613 pp. Sheehy and its supplement are authoritative guides to standard reference materials (bibliographies, indexes, dictionaries, handbooks, guides, etc.) in all subjects and fields. Relevant classified annotated entries under "Aeronautical and space engineering," "Electrical and electronics engineering," "Communications," and elsewhere. Author and title index. Sheehy is particularly useful for advanced researchers attempting comprehensive literature searches.

See 0363. Haynes, **INFORMATION SOURCES IN INFORMATION TECHNOLOGY.**

See 0369. Powell, **SELECTIVE GUIDE TO LITERATURE ON TELECOMMUNICATIONS.**

0005   Sova, Harry W., and Patricia L. Sova. **COMMUNICATION SERIALS.** Virginia Beach, VA: SovaComm, 1992, 1,041 pp. Subtitle: "International guide to periodicals in communication, popular culture, and the performing arts." Detailed entries for English language current and ceased journals include notes on contents, with lists of typical features; valuable lists of indexes for each title. Indexes for abstracts and indexes (listing journals covered), associations, columnists, departments (useful for identifying titles with regular features under "Technology" and other "tele-"words), country of publication, ISSN, publishers, titles, variant titles, subtitles, and subjects. Relevant titles under "Cable Television," "Telecommunication," as well as in subdivision "Engineering" under other main headings, like "Radio."

0006   Sterling, Christopher H. **TELECOMMUNICATIONS, ELECTRONIC MEDIA, AND GLOBAL COMMUNICATIONS: A SURVEY BIBLIOGRAPHY.** Annandale, VA.: Communications Booknotes, 1994, 47 pp. About 500 references (mostly after 1990) in subject and format arrangement, with further topical subdivisions: dictionaries, encyclopedias, glossaries; directories and yearbooks; bibliographies; development; technology; industry, economics; content; impact, effects; policy and regulation; area studies. Concise critical annotations. Cumulates and updates Sterling's previous separate bibliographies formerly issued about every two years (last revised in 1991): FOREIGN AND INTERNATIONAL COMMUNICATIONS SYSTEMS: A SURVEY BIBLIOGRAPHY; BIBLIOGRAPHY OF MASS COMMUNICATION AND ELECTRONIC MEDIA; and TELECOMMUNICATIONS POLICY: A SURVEY BIBLIOGRAPHY.

0007   **ULRICH'S INTERNATIONAL PERIODICALS DIRECTORY.** New York: Bowker, 1932–date, annual. Online version available from Dialog: covers current edition, updated monthly. CD-ROM version available from Bowker: covers current edition, updated quarterly. Standard guide to journals published worldwide. Classified arrangement includes brief entries that identify subscription information, first publication dates, editors and editorial offices and telephone and fax numbers, and indexing, with occasional descriptive notes on contents. Journals in telecommunications listed in chapters for "Communications," "Computers," "Engineering—Electrical Engineering," and others. Title keyword index.

0008   Whitaker, Marian, Ian Miles, with John Bessant and Howard Rush. **BIBLIOGRAPHY OF INFORMATION TECHNOLOGY: AN ANNOTATED CRITICAL BIBLIOGRAPHY OF ENGLISH LANGUAGE SOURCES SINCE 1980.** Aldershot: Edward Elgar; Brookfield, VT: Gower, 1989, 313 pp. "Over 500 annotated references" for books, articles, occasional papers, quasi-governmental documents (UN, OECD) "on information technology and society" since 1980 (p. 1). Arranged in 10 chapters, with subdivisions, including 1. Technology, 1.4. Telecommunications; 2. Social Trends; 3. International Economy and Politics (covers international competition and trade policy, transborder information flow); 4. Employment Debate; 5. Quality of Working Life and Work Organization (covers health and safety, new work patterns and working time); 6. Industrial Structures; 7. Organizational Issues; 8. Household and Community, 8.2. Teleservices; 9. Community and Politics; 10. Bibliographies and Indexes. Chapters include introductory essays that identify major themes and trends. Very useful author and subject indexes: relevant headings include "distance learning," "telebanking," "telecommunications," "teleconferencing," and others.

1-B. Selected Abstracting, Indexing, and Electronic Database Services

See 0490. **ABI/INFORM.**

0009   **ACCESS: THE SUPPLEMENTARY INDEX TO PERIODICALS.** Birmingham, AL: John Gordon Burke, 1975–date, 3/year. Available on disk from Electronic Information Services: covers pre-publication issues only. Intends to complement READERS' GUIDE (0032) in journal and subject coverage. Author and subject indexes for local and regional magazines as well as such journals as WIRED, INC., and WORTH. Useful for local public relations pieces about telecommunication and electronics companies.

See 0372. **ACM GUIDE TO COMPUTING LITERATURE.**

See 0071. **AMERICA: HISTORY AND LIFE.**

See 0373. **APPLIED SCIENCE AND TECHNOLOGY INDEX.**

0010   **ARTICLE 1ST.** [Online service.] Columbus, OH: OCLC, 1992–date, updated daily. Covers 1990–date. Offers keyword and selected field searching of "over 9,200 journals in science, technology, medicine, social science, business, the humanities, and popular culture," with OCLC holdings. A subject search on "telecommunication(s)" identifies nearly 2,000 articles. Journal title searching provided by CONTENTS 1ST (see 0018).

0011   **BIBLIOGRAPHIC INDEX.** New York: H. W. Wilson, 1938–date, 3/year. Online version available from Wilsonline: covers November 1984–date, updated twice weekly. Offers author and subject access to bibliographies published separately as books and articles and as parts of books. Selectively indexes about 2,600 journals and book-length bibliographies listed in CUMULATIVE BOOK INDEX (0017).

0012   **BIOGRAPHY AND GENEALOGY MASTER INDEX.** Detroit: Gale, 1980 (2nd ed.), 8 vols. **SUPPLEMENT.** 1981/82–date, annual, with five-year cumulations. Online version BIOGRAPHY MASTER INDEX available from Dialog: cumulates all editions and supplements, updated annually. CD-ROM version available from Gale: cumulates all editions, updated annually. Cross references more than 4 million biographies published in over 400 current and retrospective biographical dictionaries and other sources, including DICTIONARY OF AMERICAN BIOGRAPHY, WHO'S WHO IN AMERICA, AMERICAN MEN AND WOMEN OF SCIENCE, DICTIONARY OF SCIENTIFIC BIOGRAPHY (0097), and BIOGRAPHY INDEX (0013), as well as less well-known compilations, such as Cortada's HISTORICAL DICTIONARY OF DATA PROCESSING BIOGRAPHIES (0288) and Robert Slater's PORTRAITS IN SILICON (Cambridge, MA: MIT Press, 1987). Most useful for obscure individuals.

0013   **BIOGRAPHY INDEX: A QUARTERLY INDEX TO BIOGRAPHICAL MATERIAL IN BOOKS AND MAGAZINES.** New York: H. W. Wilson, 1946–date, quarterly, annual, and three-year cumulations. Online version available from Wilsonline, OCLC (BIOGRAPHYIND): covers August 1984–date, updated twice weekly. CD-ROM version available from H. W. Wilson: covers 1984–date, updated quarterly. Identifies biographical materials published as books, book chapters, and journal articles. Based on publications covered in other H.W. Wilson indexes (0021, 0023, 0032, 0650). "Index of Professions and Occupations" offers additional access for engineers, CEOs, etc.

0014   **BOOK REVIEW DIGEST.** New York: H. W. Wilson, 1905–date, monthly, annual cumulations. Online version available from Wilsonline, OCLC (BOOKREVDIGST): covers January 1983–date, updated twice weekly. CD-ROM version available from H. W. Wilson: covers 1983–date, updated quarterly. Lists and excerpts reviews of books by subject. Important for historical coverage. Based on publications covered in other H. W. Wilson indexes (0021, 0023, 0032, 0650).

0015   **BOOK REVIEW INDEX.** Detroit: Gale, 1965–date, bimonthly. Online version available from Dialog: covers 1969–date, updated 3/year. Complements coverage of reviews in BOOK REVIEW DIGEST (0014), indexing nearly 500 journals including BYTE, COMPUTERS AND THE HUMANITIES, and TECHNOLOGY AND CULTURE. Electronic version allows useful key word in title access.

See 0495. **BUSINESSORGS.**

See 0492. **BUSINESS PERIODICALS INDEX.**

See 0739. **CIS INDEX TO PUBLICATIONS OF THE UNITED STATES CONGRESS.**

0016   **COMMUNICATION ABSTRACTS.** Thousand Oaks, CA: Sage, 1978–date, 6/year. The most important index for communication field's scholarly literature, including titles in management and economics, policy and law, and research. International coverage limited to English-language sources. Issues abstract about 50 books and book chapters and articles from about 250 journals, with annual cumulative author and subject indexes. Main weaknesses are its relatively brief publication history, inflexible subject headings, and coverage of only the most significant telecommunication journals. Not available in electronic form.

See 0930. **COMMUNICATION CONTENTS SISALLOT.**

See 0375. **COMPUTER AND CONTROL ABSTRACTS.**

0017   **CUMULATIVE BOOK INDEX.** New York: H.W. Wilson, 1898–date, monthly. Online version available from Wilsonline: covers 1982–date, updated weekly. CD-ROM version available from Wilsondisc: covers 1982–date, updated quarterly. Gives bibliographic data for new English language books published worldwide. Useful for identifying new books on telecommunications: subject headings include many "tele-"words.

0018   **CONTENTS 1ST.** [Online service]. Columbus, OH: OCLC, 1992–date, updated daily. Covers 1990–date. Offers "complete table of contents page and holdings information for more than 9,200 journals." Covers major telecommunications trade and scholarly journals. CONTENTS 1ST and its companion service, ARTICLE 1ST (0010), which offers subject and keyword searching of the same journals, combine features of traditional indexes and current contents services with OCLC's bibliographic holdings information.

0019   **CURRENT CONTENTS.** Philadelphia: Institute for Scientific Information, 1961–date, weekly. Online version available from BRS, Dialog: covers current 6 months, updated weekly. Series of field- and discipline-specific services that reprint tables of contents of recent issues from a pool of about 6,600 journals and of selected "new, multi-authored books" analysed in ISI's citation indexes (0381, 0649), thereby including telecommunication's core journals. "Title Word" indexes conclude issues (modeled on the citation indexes' "Permuterm Index"), alphabetically listing "significant words in every article and book title." Relevant CURRENT CONTENTS services include:

    1 **CURRENT CONTENTS: ARTS & HUMANITIES.** 1978–date, weekly.

    2 **CURRENT CONTENTS: ENGINEERING, TECHNOLOGY & APPLIED SCIENCES.** 1970–date, weekly.

    3 **CURRENT CONTENTS: SOCIAL AND BEHAVIORAL SCIENCES.** 1961–date, weekly.

See 0645. **CURRENT INDEX TO JOURNALS IN EDUCATION.**

See 0545. **DISCLOSURE.**

0020   **DISSERTATION ABSTRACTS INTERNATIONAL.** Ann Arbor, MI: University Microfilms, 1938–date, monthly. Online version available from BRS, Dialog, Knowledge Index, Tech Data: covers 1861–date, updated monthly. CD-ROM version DISSERTATION ABSTRACTS ONDISC available from University Microfilms International: covers 1861–date, updated quarterly. Identifies academic research in all aspects of telecommunication. Monthly listing of doctoral dissertations, with abstracts of up to 350 words by the candidates, deposited for microfilming and sale by University Microfilms International. Section A (for humanities and social sciences) includes "Information Science"; section B (sciences and engineering) includes physics ("Electronics and Electricity"), engineering ("Electronics and Electrical"), and computer science. Issues include separate author and title word indexes. Cumulated in COMPREHENSIVE DISSERTATION INDEX, 1861–1972 (Ann Arbor, MI: Xerox University Microfilms, 1973), with annual volumes (Ann Arbor, MI: Xerox University Microfilms, 1974–date), arranged by subject discipline with key-word and cumulative author indexes. Electronic versions on BRS and Dialog cumulate DAI, AMERICAN DOCTORAL DISSERTATIONS, COMPREHENSIVE DISSERTATION INDEX, and MASTERS ABSTRACTS.

See 0376. **ELECTRICAL AND ELECTRONICS ABSTRACTS.**

See 0377. **ELECTRONICS AND COMMUNICATIONS ABSTRACTS JOURNAL.**

See 0378. **ENGINEERING INDEX.**

0021　**ESSAY AND GENERAL LITERATURE INDEX.** New York: H. W. Wilson, 1934–date, semiannual with annual and irregular cumulations. Online version available from Wilsonline: covers January 1985–date, updated twice weekly. CD-ROM version available from H. W. Wilson: covers 1985–date, updated quarterly. Relevant to all areas of telecommunication. Indexes book chapters or parts of collections, such as published proceedings or symposia papers, in all disciplines. Subject headings include "Computer vision," "Optoelectrical devices," "Technological innovations," and "Telecommunications." Cumulative index to nearly 10,000 analyzed collections: ESSAY AND GENERAL LITERATURE INDEX: WORKS INDEXED 1900–1969 (1972).

See 0493. **F & S INDEX UNITED STATES.**

0022　**GENERAL SCIENCE INDEX.** New York: H. W. Wilson, 1978–date, monthly. Online version available from Wilsonline, OCLC (GENSCIINDEX): covers May 1984–date, updated twice weekly. CD-ROM version available from H. W. Wilson, Silver Platter: covers 1984–date, updated quarterly. Indexes telecommunication's most readily available science and technology journals, as well as major science and technology journals that have frequently published articles on telecommunication, such as ISSUES IN SCIENCE AND TECHNOLOGY, NATURE, NEW SCIENTIST, and TECHNOLOGY REVIEW. Not a substitute for comprehensiveness of ENGINEERING INDEX (0378) or APPLIED SCIENCE AND TECHNOLOGY INDEX (0373).

See 0379. **GOVERNMENT REPORTS ANNOUNCEMENTS AND INDEX.**

0023　**HUMANITIES INDEX.** New York: H. W. Wilson, 1974–date, quarterly with annual cumulations. Online version available from Wilsonline, OCLC (HUMANITIESIN): covers February 1984–date, updated twice weekly. CD-ROM version available from H. W. Wilson, Silver Platter: covers 1984–date, updated quarterly. Useful for identifying scholarly studies of telecommunication's impact on society and culture, both historically and currently. Useful headings include "Automatic speech recognition," "Electronic data publishing," "Facsimile transmission," "High definition television," "Teleconferencing," among others.

See 0401. **IHS INTERNATIONAL STANDARDS & SPECIFICATIONS.**

See 0494. **INDEX OF ECONOMIC ARTICLES.**

See 0380. **INDEX TO IEEE PUBLICATIONS.**

See 0726. **INDEX TO LEGAL PERIODICALS.**

See 0084. **INDEX TO NIDS.**

See 0932. **INDICATOR CONTENTS.**

0024　**INFORMATION SCIENCE ABSTRACTS.** New York: Plenum, 1966–date, monthly. Formerly DOCUMENTATION SCIENCE ABSTRACTS (1966–1968) and other variant titles. Online version available from Dialog: covers 1966–date, updated monthly. Useful coverage of information science and technology literature, including books, conference proceedings, reports, patents, and major telecommunication and information technology journals. Classified arrangement provides relevant listings under 3.6. Technology transfer; 3.11. Communications and telecommunications systems; 3.12. Radio, television, video; and 3.13. Office communication and automation. Monthly and cumulative annual author and subject indexes. Useful headings include "Telecommunications," "Telecomuting," "Telemedicine," "Teleconferencing," "Telephone," "Telepresence," "Teleshopping," among others.

0025　**JOURNALISM ABSTRACTS.** Columbia, SC: Association for Education in Journalism and Mass Communication, 1963–date, annual. Indexes and abstracts Ph.D. dissertations and master's theses in journalism and mass communication from universities in U.S. and Canada, with abstracts written by the degree candidates. Indexes for authors, degree-granting institutions, and subjects, with headings derived from key words in titles and abstracts, including proper names. Useful headings include "Satellite communication" and "Telephones."

See 0727. **LEGAL TRAC.**

See 0728. **LEXIS/NEXIS.**

0026　**LIBRARY AND INFORMATION SCIENCE ABSTRACTS.** London: Library Association, 1950–date, monthly. Formerly LIBRARY SCIENCE ABSTRACTS (1950–1968). Online version LISA available on Dialog, BRS: cover 1967–date, updated monthly. CD-ROM version LISA-PLUS available from Bowker Electronic Publishing: covers 1969–date, updated quarterly. International bibliography with abstracts of book and library science, information science, and related disci-, plines. Very strong coverage of new communication and information technologies, including teleconferencing, videotex, databases and online systems, and electronic publishing. Indexes over 500 journals.

0027　**LIBRARY LITERATURE: AN INDEX TO LIBRARY AND INFORMATION SCIENCE.** New York: H. W. Wilson, 1934–date, bimonthly. Covers from 1921–date. Online version available from Wilsonline, OCLC (LIBRARYLIT): covers 1984–date, updating varies. CD-ROM version available from H. W. Wilson: covers 1984–date, updated quarterly. International coverage of books, chapters, and articles in 250 journals, including major information science and technology titles. Useful for articles on the Internet and other data communication networks, U.S. government information agencies (NTIS, NSF, Office of Technology Assessment), and new communication technologies (teleconferencing, facsimile transmission, CD-ROMs).

See 0647. **MANAGEMENT CONTENTS.**

See 0934. **MASS COM PERIODICAL LITERATURE INDEX.**

0028　**MICROCOMPUTER ABSTRACTS.** [Online service]. Columbus, OH: OCLC, 1994–date, updated monthly. Covers January 1989–date. Database produced by Learned Information, Inc. Subject, product, and company name, and other indexing of "more than 75 popular magazines and professional journals on microcomputing in business, education, industry, and the home," with abstracts. Covers INFOWORLD, LINK-UP, PC MAGAZINE, BYTE, COMPUTE, and others. Search on "telecommunication(s)" produces over 2,000 "hits."

See 0734. **MONTHLY CATALOG OF UNITED STATES GOVERNMENT PUBLICATIONS.**

0029　**NEWSPAPER ABSTRACTS** (NEWSABS). [Online service]. Columbus, OH: OCLC, 1992–date, updated weekly. Covers 1989–date. CD-ROM version available from UMI/Data Courier. Indexes and abstracts 25 selected major national and regional newspapers, including NEW YORK TIMES, WALL STREET JOURNAL, WASHINGTON POST, and others. Useful service for current awareness, but LEXIS/NEXIS (0728) offers wider coverage of local and regional newspapers.

0030　**PAIS BULLETIN.** New York: Public Affairs Information Service, 1914–date, monthly. Online version available from BRS, Dialog, Knowledge Index, OCLC (PAIS DECADE), Tech Data: coverage varies (Dialog 1972–date), updated monthly. CD-ROM version available from Silver Platter: covers 1972–date, updated quarterly. Subject indexing of international scholarly and professional publications (monographs, 1,400 journals, U.S. and foreign governmental, quasi-governmental, and organizational documents) in worldwide English-language public policy literature—law, political science, economic and social conditions, public administration, and international relations. Covers core telecommunication journals, including GOVERNMENT INFORMATION QUARTERLY (1114), INFORMATION SOCIETY (1104), PUBLIC UTILITIES FORTNIGHTLY (1119), SPACE POLICY (1120), TELECOMMUNICATIONS POLICY (1123), and TDR: TRANSNATIONAL DATA AND COMMUNICATIONS REPORT (1151); monographs from major telecommunication publishers, like Artech, Erlbaum, Computer Science Press, Sage, and Wiley; and publications from the FCC, ITU, UNESCO, OECD, and NTIS. Annual cumulative author index. Useful subject headings include "Communication Systems" (used for telecommunications), "Remote Sensing Systems," "Optical Storage Devices," "Radio Telephones," "Electronic Mail," "Electronic Publishing," "Satellites—Communication Uses," "Teletext," "Teleconferencing," "Videotex." PAIS FOREIGN LANGUAGE INDEX (1971–date) provides useful complementary coverage of non-English language public policy literature. Electronic versions cumulate PAIS BULLETIN and PAIS FOREIGN LANGUAGE INDEX.

0031 **PERIODICAL ABSTRACTS** (PERABS). [Online service]. Columbus, OH: OCLC, 1992– date, updated weekly. Adapted from UMI/Data Courier's PERIODICAL ABSTRACTS online database and CD-ROM (1987–date). Indexes and abstracts "1,500+ general and academic journals" in all areas. This amounts to UMI's equivalent to H.W. Wilson's READERS' GUIDE (0032), covering popular journals like TIME, BUSINESS WEEK, and VARIETY, as well as journals particularly useful for telecommunications research, including FEDERAL COMMUNICATIONS LAW JOURNAL (1113), and TELECOMMUNICATIONS POLICY (1123).

See 0648. **PSYCHOLOGICAL ABSTRACTS.**

0032 **READERS' GUIDE TO PERIODICAL LITERATURE.** New York: H. W. Wilson, 1900– date, monthly. Online version available from OCLC, Wilsonline: covers January 1983–date, updated twice weekly. Online READGUIDEABS (READERS GUIDE ABSTRACTS) also available from OCLC. CD-ROM version READERS' GUIDE ABSTRACTS available from H. W. Wilson, Silver Platter: covers 1983–date, update quarterly. Useful for identifying articles on telecommunication and new communication technologies in popular journals, such as TIME, NEWSWEEK, U.S. NEWS AND WORLD REPORT, and BUSINESS WEEK. Also indexes special interest journals that feature state-of-the-art information on telecommunication, including AVIATION WEEK & SPACE TECHNOLOGY, HIGH TECHNOLOGY BUSINESS (1090), RADIO-ELECTRONICS, SCIENCE, SCIENTIFIC AMERICAN, and TECHNOLOGY REVIEW. Useful subject headings include "Cellular Radio," "Compact Discs," "Digital Audio Tape Recorders and Recording," "Direct Broadcast Satellite Services," "Electronic Mail Systems," "Fax Machines," "High Definition Television," "Local Area Networks," and "Teleconferencing."

See 0381. **SCIENCE CITATION INDEX.**

See 0649. **SOCIAL SCIENCE CITATION INDEX.**

See 0650. **SOCIAL SCIENCES INDEX.**

See 0651. **SOCIOLOGICAL ABSTRACTS.**

See 0513. **STATISTICAL MASTERFILE.**

See 0937. **TRANSDEX INDEX.**

See 0953. **UNDOC: CURRENT INDEX: UNITED NATIONS DOCUMENTS INDEX.**

See 0384. U.S. Department of Commerce. Patent and Trademark Office. **CASSIS.**

See 0729. **WESTLAW.**

1-C. Directories and Yearbooks

0033 **AMERICAN LIBRARY DIRECTORY: A CLASSIFIED LIST OF LIBRARIES IN THE UNITED STATES AND CANADA.** New York: R. R. Bowker, 1923–date, annual. CD-ROM version available from Bowker. A standard reference source for addresses, phone numbers, personnel, and brief information on American libraries. Information arranged by state for all varieties of libraries (corporate, state, academic, research, etc.), with brief information on special collections.

0034 **ANNUAL REVIEW OF INFORMATION SCIENCE AND TECHNOLOGY.** Medford, NJ: Learned Information, 1966–date, annual. Sponsored by American Society for Information Science. Extensive literature review articles on developments in information technology. Regular departments cover planning information systems and services, techniques and technologies, the infor- mation profession. Recent volumes address information technology for distance learning, information pricing, "Information Technology and the Individual." Comprehensive author and keyword indexes to all volumes. Relevant headings include "Communications," "Computers," "Technology," "Telecommunications."

0035   **ENCYCLOPEDIA OF ASSOCIATIONS**. Detroit: Gale, 1956–date, annual. Online version available from Dialog: covers current edition, updated semiannually. CD-ROM version GALE GLOBAL DIRECTORY available from Gale: covers current edition, updated semiannually. 28th ed. for 1994. Standard directory describing "nearly 23,000 national and international organizations" in all fields and of all kinds. Classified listings relevant to telecommunications appear in all chapters, especially those for trade and business; labor unions; engineering and technology professions; and public affairs. Careful use of "Name and Keyword Index" is essential: listings under "Cable," "Satellites," "Telecommunications," "Telephones," and others. Also includes geographic and executive indexes. Electronic versions cumulate entries for some 87,000 organizations listed in Gale's ENCYCLOPEDIA OF ASSOCIATIONS, INTERNATIONAL ORGANIZATIONS (1984–date), and REGIONAL, STATE, AND LOCAL ORGANIZATIONS (1987–date).

0036   **EVENTLINE**. [Online service]. Columbus, OH: OCLC, 1992–date, updated monthly. Database produced by Elsevier Science Publishers. Offers subject, event, location, and other indexing for international "conventions, conferences, symposia, trade fairs, and exhibits scheduled between the present time and the 21st century." Entries give data for contacts, sponsoring organizations, exhibits (telephone, fax numbers, etc.). Useful supplement to calendars in professional and trade journals.

0037   **GALE DIRECTORY OF DATABASES**. Kathleen Young Maraccio, editor. Detroit: Gale, 1979–date, annual. Formerly DATABASE OF DATABASES, COMPUTER-READABLE DATABASES, and CUADRA DIRECTORY OF DATABASES. Online version available from Dialog, Orbit, Data-Star: covers current edition, updated semiannually. CD-ROM version available from Silver Platter: covers current edition, updated semiannually. Directory describing electronic resources. Volume 1 lists over 5,000 online databases; volume 2 lists CD ROM, floppy disk, and tape products. Indexed by subject, producer, distributor, country; "Master Index" includes product and organization names.

0038   Krol, James, editor. **TELECOMMUNICATIONS DIRECTORY**. Detroit: Gale, 1983–date, biennial. 6th ed. for 1994/1995 (1993). Guide to about "2,300 national and international voice and data communications networks, electronic mail services, teleconferencing facilities and services, facsimile services, videotex and teletext operations, transactional services, local area networks, audiotex services, microwave systems/networkers, satellite facilities, and others involved in telecommunications, including related consultants, advertisers/marketers, associations, regulatory bodies, organizations and services in all areas of telecommunications" (subtitle). All-purpose guide.

0039   **RESEARCH CENTERS DIRECTORY**. Detroit: Gale, 1960–date, annual. 19th ed. for 1995 (1994). Online version available from Dialog: covers current edition, updated semiannually. Describes more than 13,000 university-related and not-for-profit research organizations. Topical arrangement: relevant centers identified in sections "Astronomy and Space Sciences," "Computers and Mathematics," "Engineering and Technology," and others. Subject, geographic, and master (keyword) indexes. Electronic version cumulates listings in RESEARCH CENTERS DIRECTORY, INTERNATIONAL RESEARCH CENTERS DIRECTORY (1982–date), and GOVERNMENT RESEARCH DIRECTORY (0732).

1-D. Dictionaries

0040   **COMPUTER DICTIONARY: THE COMPREHENSIVE STANDARD FOR BUSINESS, SCHOOL, LIBRARY, AND HOME**. Redmond, WA: Microsoft Press, 1994 (2nd ed.), 442 pp. Computer terms and acronyms from variety of fields, with illustrations and tables.

0041   Freedman, Alan. **THE COMPUTER GLOSSARY: THE COMPLETE ILLUSTRATED DESK REFERENCE**. New York: AMA-COM, 1993 (6th ed.), 574 pp. 3,500 terms with illustrations and drawings.

0042 **GLOSSARY OF TELECOMMUNICATIONS TERMS: ENGLISH, ARABIC, FRENCH, SPANISH**. Geneva, Switzerland: United Nations Development Programme; Arab Fund for Economic and Social Development, 1987, 1,005 pp. Polygot dictionary.

0043 Graham, John. **THE FACTS ON FILE DICTIONARY OF TELECOMMUNICATIONS**. New York: Facts on File, 1983, 199 pp. Some 2,000 items.

0044 Hansen, Douglas E. **EDUCATIONAL TECHNOLOGY TELECOMMUNICATIONS DICTIONARY WITH ACRONYMS**. Englewood Cliffs, NJ: Educational Technology Publications, 1991, 55 pp. Non-technical definitions of about 750 common terms and acronyms; intended for educational administrators and teachers as well as business and community leaders.

0045 Held, Gilbert. **DATA AND COMPUTER COMMUNICATIONS: TERMS, DEFINITIONS, AND ABBREVIATIONS**. New York: John Wiley, 1989, 254 pp. Some 7,000 terms.

0046 Keen, Peter G. W. **EVERY MANAGER'S GUIDE TO INFORMATION TECHNOLOGY: A GLOSSARY OF KEY TERMS AND CONCEPTS FOR TODAY'S BUSINESS LEADER**. Cambridge: Harvard Business School Press, 1991, 170 pp. Some 140 terms are discussed at length and with wit.

0047 Kleczek, Josip. **SPACE SCIENCES DICTIONARY**. Amsterdam: Elsevier Science Publishing, 1990–date, in progress; vol. 3 covering "Space Technology/Space Research" published in 1994. English, French, German, Spanish, Portuguese, Russian.

0048 Kurpis, Gediminas P. **THE NEW IEEE STANDARD DICTIONARY OF ELECTRICAL AND ELECTRONICS TERMS**. New York: IEEE, 1993, 1,619 pp. Defines technical terms and abstracts IEEE standards.

0049 Langley, Graham. **TELEPHONY'S DICTIONARY: DEFINING 16,000 TELECOMMUNICATIONS WORDS AND TERMS**. Chicago: Telephony, 1986 (2nd ed.), 402 pp. Title sums it up—more technical than others on this list.

0050 Lauriston, Andy, and Jocelyne Le Neal. **BILINGUAL DICTIONARY OF INTERNATIONAL TELECOMMUNICATIONS**. Montreal: Teleglobe Canada, 1983–1988, 3 vols. One each on Antenna Theory, Transmission Equipment, and Switching Equipment. Each provides the French and English definitions, with diagrams.

0051 Longley, Dennis, and Michael Shain. **VAN NOSTRAND REINHOLD DICTIONARY OF INFORMATION TECHNOLOGY**. New York: Oxford, 1989 (3rd ed.), 566 pp. Terms related to computer networks, data communication, cryptography, expert systems, expert systems, microcomputers, online information services, and more.

0052 Newton, Harry. **NEWTON'S TELECOM DICTIONARY** New York: Flatiron Publishing, 1994, 1,177 pp. Larger print makes for easier reading of these thousands of definitions.

0053 Parker, Sybil P. ed. **COMMUNICATIONS SOURCE BOOK**. New York: McGraw-Hill, 1989, 359 pp. Originally appeared in McGRAW-HILL ENCYCLOPEDIA OF SCIENCE AND TECHNOLOGY, 6th ed. (1987). Includes chapters on signal transmission and network and switching.

0054 Ralston, Anthony, and Edwin D. Reilly, eds. **ENCYCLOPEDIA OF COMPUTER SCIENCE AND ENGINEERING**. New York: Van Nostrand Reinhold, 1993 (3rd ed.), 1,558 pp. Over 600 authoritative articles on computer science and information technology. Appendices list major journals, computer science graduate programs, high-level computer languages, polygot glossary, and chronology.

0055 Rosenberg, Jerry M. **DICTIONARY OF COMPUTERS, DATA PROCESSING AND TELECOMMUNICATIONS**. New York: Wiley, 1987 (2nd ed.), 734 pp. Focus is on the former in this revision of a 1984 standard guide.

0056 Spencer, Donald D. **COMPUTER DICTIONARY**. 4th ed. Ormond Beach, FL: Camelot Publishing, 1993, 459 pp. Terms, with portraits of computer pioneers.

0057   Towell, Julie E., and Helen E. Sheppard. **COMPUTER & TELECOMMUNICATIONS ACRONYMS.** Detroit: Gale Research, 1986, 391 pp. More than 25,000 entries, including technical terms and associations.

0058   Weik, Martin H. **COMMUNICATIONS STANDARD DICTIONARY.** New York: Van Nostrand Reinhold, 1989 (2nd ed.), 1,219 pp. Comprehensive collection of technical terms related to information transfer, with illustrations and tables. Emphasis on standards. Well cross-referenced.

0059   _____. **FIBER OPTICS STANDARD DICTIONARY.** New York: Van Nostrand Reinhold, 1989 (2nd ed.), 366 pp. Comprehensive collection of technical terms. Well cross-referenced.

1-E. Encyclopedias

0060   Froehlich, Fritz, and Allen Kent, editors. **THE FROEHLICH/KENT ENCYCLOPEDIA OF TELECOMMUNICATIONS.** New York: Marcel Dekker, 1991–date (in progress). An ambitious attempt to assemble authoritative articles on all aspects of telecommunications. Entries offer basic overviews of particular technologies as well as broader surveys of less well-defined topics, such as the "Divestiture of AT&T." Other entries cover historical figures, like Ezra Cornell and A.E. Dolbear, companies and organizations, systems, events in telecommunications history, and the like. Projected completion in 10 volumes seems optimistic. Volumes that will cover Morse and the many "tele"-words have yet to be produced; supplements to cover omissions will be needed.

0061   Meyers, Robert A., editor. **ENCYLOPEDIA OF TELECOMMUNICATIONS.** New York: Academic Press, 1989, 526 pp. Some 28 chapters by more than 40 authorities and experts (scholars and professionals) cover theoretical, political, economic, and, in particular, technical topics in telecommunications. Meyers indicates that the essays intend to survey telecommunications' "three major components—transmission, the communications channel, and reception, providing basic theory, derivations of mathematical relationships, hardware, history, status, and a forecast of future directions" (p. xi). Several essays cover telecommunications in general, most notably Herbert Dordick's "Telecommunications" (pp. 455–76)—a good starting point. Essays include glossaries, bibliographies, and illustrations. Index.

History

Chapter 2 begins with several standard bibliographies that describe telecommunication's historical literature published as books and journals. Particularly noteworthy among the bibliographies and reference resources for secondary scholarship on telecommunications history is the ISIS CUMULATIVE BIBLIOGRAPHY (0067). This chapter goes on to include references to significant collections of unpublished materials located in museums, archives, libraries, and other repositories. These guides clearly reflect telecommunication's obvious technological and business/labor interests. Section 2-B in fact, includes only a limited selection of federal, state, and local museums, archives, historical collections, public and academic libraries, and other repositories containing important manuscripts and unpublished materials on telecommunications. Databases described in this chapter identify many more collections. On the other hand, no existing guides describe the full range of corporate archives—and some archivists and librarians are hesitant to even admit their existance. We apologize for any omissions and welcome additional information about other significant collections. Entries for collections cite standard NATIONAL UNION CATALOG OF MANUSCRIPT COLLECTIONS (0085) numbers (MS) to facilitate obtaining more information about their contents. Users are advised to consult DeWitt's outstanding BIBLIOGRAPHIC GUIDE TO ARCHIVES AND MANUSCRIPT COLLECTIONS IN THE UNITED STATES (0064) to identify published guides, finding aids, and calendars for repositories. DeWitt's selection of guides to the collections of the Library of Congress, National Archives, Smithsonian, and presidential libraries will save researchers valuable time. This chapter concludes with selected listings of general secondary historical studies of telecommunication technologies—the telegraph, telephone, wireless/radio, and satellite; bibliographies; industry and company histories; and international telecommunications.

2-A Bibliographic Resources

2-A-1. Bibliographies

0062 **CATALOGUE OF BOOKS AND PAPERS RELATING TO ELECTRICITY, MAGNETISM, THE ELECTRIC TELEGRAPH, &c., INCLUDING THE RONALDS LIBRARY ... WITH A BIOGRAPHICAL MEMOIR**. Francis Ronalds, compiler; Alfred J. Frost, editor. London: E. & F.N. Spon, 1880, 564 pp. "Published by the Society of Telegraph Engineers." Describes early materials.

0063 Cortada, James W. **A BIBLIOGRAPHIC GUIDE TO THE HISTORY OF COMPUTING, COMPUTERS, AND THE INFORMATION PROCESSING INDUSTRY**. Westport, CT: Greenwood, 1990, 656 pp. More than 4,500 items with both author and subject indexes.

0064 DeWitt, Donald L. **GUIDE TO ARCHIVES AND MANUSCRIPT COLLECTIONS IN THE UNITED STATES: AN ANNOTATED BIBLIOGRAPHY**. Westport, CT: Greenwood, 1994, 478 pp. Describes 2,062 published finding aids in classified arrangement. Identifies finding aids for business collections, federal archives, and other collections. Very useful list of guides and aids for electrical engineering and electronic technology collections (pp. 199–200). An important starting point for historical research.

0065   Finn, Bernard S. **THE HISTORY OF ELECTRICAL TECHNOLOGY: AN ANNOTATED BIBLIOGRAPHY.** New York: Garland, 1991, 360 pp. Some 1,500 unannotated items with both author and subject indexes. Of most interest for its first two sections: broad historical works, and communications, each of which is further divided into sub-sections. Index.

0066   Gluckman, Albert Gerard. **THE INVENTION AND EVOLUTION OF THE ELECTRO-TECHNOLOGY TO TRANSMIT ELECTRICAL SIGNALS WITHOUT WIRES: AN ANNOTATED BIBLIOGRAPHY OF 17th, 18th, AND 19th CENTURY EXPERIMENTAL STUDIES OF ELECTROSTATIC INDUCTION, SPARK-GAP AND LIGHTNING DISCHARGES, MAGNETIC INDUCTION, OSCILLATING CIRCUITS, RESONANCE, AND ELECTROMAGNETIC WAVE PROPAGATION.** Washington: Washington Academy of Sciences, Mount Vernon College, 1993, 239 pp. Useful for information on historical basic research on wireless communications. Describes 311 books, articles, and reports. Arranged in reverse chronological order from 1922–1664 "in order to project the appearance of looking backward into time from the viewpoint of electromagnetic science at the turn of the 20th century" (p. 14). Includes chronology of "evolutionary milestones," name index.

See 0294. **INTERNATIONAL DIRECTORY OF COMPANY HISTORIES.**

0067   **ISIS CUMULATIVE BIBLIOGRAPHY: A BIBLIOGRAPHY OF THE HISTORY OF SCIENCE FORMED FROM ISIS CRITICAL BIBLIOGRAPHIES 1–110, 1913–1985.** London: Mansell; Boston: G.K. Hall, 1971, 1980, 1989, 10 vols. Online version available from Research Libraries Group. Published "in conjunction with the History of Science Society." Three separately published similarly organized sets that cover the world's history of science and technology literature. Classified annotated listings for (1) personalities and institutions (Morse, Bell, Edison, Bell Telephone Laboratories, IEEE); and (2) general subjects and subjects in historical periods. Communications included in classifications for "X," with subdivisions. Indexes for book reviews, authors, and subjects: relevant listings under "Communication technology," "Electrical communications," etc. Updated by bibliographies featured in TECHNOLOGY AND CULTURE (1042).

0068   Mottelay, Paul Fleury. **BIBLIOGRAPHICAL HISTORY OF ELECTRICITY AND MAGNETISM CHRONOLOGICALLY ARRANGED.** London: Griffin, 1922, 673 pp. Complements catalogues of Weaver (0070) and Ronalds (0062) for early materials.

0069   Shiers, George, assisted by May Shiers. **BIBLIOGRAPHY OF THE HISTORY OF ELECTRONICS** Metuchen, NJ: Scarecrow Press, 1972, 323 pp. Invaluable guide primarily devoted to communications, with sections on radio, telegraphy and telephony, and television and facsimile as well as more general resources, such as biographies and company histories. Some 1,800 annotated entries, including articles featured in early telecommunication and electronics journals. Author and subject indexes.

0070   Weaver, William D. **CATALOGUE OF THE WHEELER GIFT OF BOOKS, PAMPHLETS, AND PERIODICALS TO THE LIBRARY OF THE AMERICAN INSTITUTE OF ELECTRICAL ENGINEERS.** New York: American Institute of Electrical Engineers, 1909 (2 vols), 504 + 475 pp. The standard descriptive listing of telecommunications and electronics' early technical literature. About 6,000 concisely annotated references to both books and articles, with index.

2-A-2. Selected Abstracting, Indexing, and Electronic Database Services

0071   **AMERICA: HISTORY AND LIFE.** Santa Barbara, CA: ABC-Clio, 1964–date, 5/year. Online version available from Dialog, Knowledge Index. CD-ROM version available from ABC-Clio: covers 1982–date, updated 3/year. The best index to the scholarly literature (books, journals, dissertations) on U.S. telecommunication and communication technology history. Selectively indexes about 2,000 journals in history, economics, law, political science, and other disciplines as well as major interdisciplinary journals, in 40 languages, including JOURNAL OF COMMUNICATION (1105), TECHNOLOGY AND CULTURE, and SCIENCE AND SOCIETY. Scholarly historical literature on international telecommunication history indexed in complementary HISTORICAL ABSTRACTS (Santa Barbara, CA: ABC-Clio, 1955–date). Several separately published selected topical bibliographies in the Clio Bibliography Series, based on ABC-Clio databases, also include listings relevant to telecommunications history: LABOR IN AMERICA: A HISTORICAL BIBLIOGRAPHY (1985); CORPORATE AMERICA: A HISTORICAL BIBLIOGRAPHY (1984).

0072 **WRITINGS ON AMERICAN HISTORY.** Washington: American Historical Association, 1902–date, annual. Formerly ANNUAL REPORT of the American Historical Association (1902–1960) and other variant titles. Standard authoritative bibliography of studies of American history. More comprehensive, but also more difficult to use, than AMERICA: HISTORY AND LIFE (0071). Classified listings: unannotated, no subject indexing, no cumulations; but covers everything, international in scope. Relevant listings under "Communications History," "History of Science and Technology"; in earliest volumes, under "Communication; transportation; public works."

2-A-3. Directories of Archives, Libraries, Museums, and Special Collections

0073 Ash, Lee, and William G. Miller, compilers. **SUBJECT COLLECTIONS: A GUIDE TO SPECIAL BOOK COLLECTIONS AND SUBJECT EMPHASES AS REPORTED BY UNIVERSITY, COLLEGE, PUBLIC, AND SPECIAL LIBRARIES AND MUSEUMS IN THE UNITED STATES AND CANADA.** New York: R. R. Bowker, 1985 (6th ed.), 2 vols. A standard guide to collections, including institutional and corporate archives; subject arrangement includes detailed listings under "Communications," "Radio," "Telecommunications," etc.

0074 Bedi, Joyce E., Ronald R. Kline, and Craig Semsel. **SOURCES IN ELECTRICAL HISTORY: ARCHIVES AND MANUSCRIPT COLLECTIONS IN U.S. REPOSITORIES.** New York: Center for the History of Electrical Engineering, 1989, 234 pp. Sponsored by the Friends of the IEEE Center. Intended to update Hounshell's 1973 guide (0082). Describes 1,008 collections in 158 repositories, mainly university archives and state historical societies. Entries arranged by collection titles, with notes on size, contents, finding aids, and data for repositories and access. Repository and subject indexes. Identifies some 26 collections including materials for Morse; 9 for the Western Union Russian telegraph extension. An important resource for historical research. "Preface" notes that future volumes issued by the Center will cover oral history, business, private, and non-U.S. collections (p. v).

0075 Bruemer, Bruce H. **RESOURCES FOR THE HISTORY OF COMPUTING: A GUIDE TO U.S. AND CANADIAN RECORDS.** Minneapolis, MN: Charles Babbage Institute, 1987, 187 pp. Describes personal papers, corporate and company records, and oral history collections in about 100 repositories, including special collections and archives of Library of Congress (0119), Smithsonian (0127), MIT (0177), University of California at Berkeley (0194), AT&T (0149) and Motorola (0156). Excludes collections in the National Archives. Subject index references materials under "communications and computers," "telecommunications," "Bell system," FCC, RCA, and others.

0076 Cortada, James W. **ARCHIVES OF DATA-PROCESSING HISTORY: A GUIDE TO MAJOR U.S. COLLECTIONS.** New York: Greenwood, 1990, 181 pp. Surveys collections on electronics in National Archives (0120), National Museum of American History (0127), Library of Congress (0119), Charles Babbage Institute, University of Minnesota (0204), MIT (0177), Harvard (0175), Unisys (0162), and others. Subject index.

0077 Danilov, Victor J. **AMERICA'S SCIENCE MUSEUMS.** New York: Greenwood, 1990, 483 pp. Classified descriptions of national, regional, local, and private museums, including the Computer Museum, National Museum of Communications (0114), U.S. Army Communications-Electronics Museum (0118), Hagley Museum and Library (0137), NASA Johnson Space Center Visitor Center, and others. Museum name index.

0078 **DIRECTORY OF ARCHIVES AND MANUSCRIPT REPOSITORIES IN THE UNITED STATES.** Phoenix, AZ: Oryx Press, 1988 (2nd ed.), 853 pp. Compiled by National

Historical Publications and Records Commission, especially useful for collections in university and state and local historical and institutional collections.

0079   **DIRECTORY OF BUSINESS ARCHIVES IN THE UNITED STATES AND CANADA.** Chicago, IL: Society of American Archivists, Business Archives Section, 1990 (4th ed.), 96 pp. Includes selected telecommunication and electronics company archives. Geographical and business-type indexes.

0080   **DIRECTORY OF HISTORICAL ORGANIZATIONS IN THE UNITED STATES AND CANADA.** Mary Bray Wheeler, editor. Nashville, TN: AASLH Press, American Association for State and Local History, 1990 (14th ed.), 1,108 pp. Arranged by state and province. Descriptions of historical organizations, with brief notes on collections. See "Major Program Areas" index under "Transportation/Industry/Technology" for relevant listings. Organizational name index.

0081   **DIRECTORY OF SPECIAL LIBRARIES AND INFORMATION CENTERS.** Detroit: Gale, 1963–date, irregular. Subject index in 1994 ed. includes international listings under "Telecommunications," "Telegraph," "Telephone," etc.; with 11th ed., updated by NEW SPECIAL LIBRARIES (1988–date).

0082   Hounshell, David A. **MANUSCRIPTS IN U.S. DEPOSITORIES RELATING TO THE HISTORY OF ELECTRICAL SCIENCE AND TECHNOLOGY.** Washington: Smithsonian Institution Press, 1973, 116 pp. Sponsored by the IEEE History Committee and the Electrical Technology Group, Society for the History of Technology. Detailed descriptions of about 250 collections, arranged under nearly 100 academic, professional, and state historical archives and libraries. Name, title, and subject index.

0083   Leab, Daniel J., and Philip P. Mason. **LABOR HISTORY ARCHIVES IN THE UNITED STATES: A GUIDE FOR RESEARCHING AND TEACHING.** Detroit: Wayne State University Press, 1992, 286 pp. Authoritative descriptions of labor collections in 40 libraries, archives, research centers. Useful descriptions of materials in the Joseph A. Beirne Memorial Archives of the Communications Workers of America's Washington headquarters (0228) as well as for CWA, IUE (International Union of Electrical, Radio, and Machine Workers), and other collections related to telecommunications.

0084   Library of Congress Manuscript Division. **NATIONAL INVENTORY OF DOCUMENTARY SOURCES IN THE UNITED STATES.** Teaneck, NJ: Chadwyck-Healey, 1983–date, irregular. Microfiche collection. The most powerful access tool for U.S. archive collections, comprising a topical index to published and unpublished finding aids for individual collections, with the full text of those finding aids. CD-ROM version, INDEX TO NIDS (Alexandria, VA: Chadwyck-Healey, 1992–date), essentially makes this key word access to collection descriptions.

0085   _____. **NATIONAL UNION CATALOG OF MANUSCRIPT COLLECTIONS.** Washington: Library of Congress, 1962–date, annual, with annual and five-year cumulative indexes. Usually refered to as NUCMC (pronounced "nuck-muck"). Essential for historical research. As of 1992, includes more than 70,000 manuscripts collections reported by nearly 1,400 libraries and other repositories since 1959. Subject index provides useful headings for the full range of "tele"-words. Access greatly facilitated with INDEX TO PERSONAL NAMES IN THE NATIONAL UNION CATALOG OF MANUSCRIPT COLLECTIONS, 1959–1984, 2 vols. (Alexandria, VA: Chadwyck-Healey, 1988); and INDEX TO SUBJECTS AND CORPORATE NAMES IN THE NATIONAL UNION CATALOG OF MANUSCRIPT COLLECTIONS, 1959–1984, 2 vols. (Alexandria, VA: Chadwyck-Healey, 1994).

0086   Mount, Ellis. **SCI-TECH ARCHIVES AND MANUSCRIPT COLLECTIONS.** Binghamton, NY: Haworth, 1989, 144 pp. Also published as SCIENCE & TECHNOLOGY LIBRARIES, vol. 9, no. 4 (Summer 1989), pp. 1–104. Articles describe collections in Edison Archives, MIT, the National Archives, and selected others in science and technology fields.

0087   **OFFICIAL MUSEUM DIRECTORY.** Washington: American Association of Museums, 1961–date, biennial. Briefly annotated listings by state with subject index.

0088 **ORAL HISTORY COLLECTIONS.** Alan M. Meckler and Ruth McMullin, editors. New York: R.R. Bowker, 1975, 344 pp. Name and subject index to oral history collections in U.S. and selected foreign centers and archives. Relevant listings under "Radio," "Popular culture," and names of inventors, etc.

0089 Public Utilities Commission Information Resource Network. **DIRECTORY OF PUBLIC UTILITY COMMISSION LIBRARIES, INFORMATION CENTERS AND OTHER NARUC MEMBER LIBRARIES.** Washington: National Association of Regulatory Utility Commissioners, 1991–date, irregular. Contacts, addresses, phone and fax numbers, with notes on contents, size, and focuses of collections and other services.

0090 Rothenberg, Marc. **HISTORY OF SCIENCE AND TECHNOLOGY IN THE UNITED STATES: A CRITICAL AND SELECTIVE BIBLIOGRAPHY.** New York: Garland, 1982, 242 pp. Some 832 classified annotated entries: chapters cover "Electricity and Electronics," "Space Exploration." Author and subject indexes; relevant listings under Bell, Bell Laboratories, Edison, Morse, telegraph, telephone, Western Union, etc.

0091 Warnow, Joan Nelson. **A SELECTION OF MANUSCRIPT COLLECTIONS AT AMERICAN REPOSITORIES.** New York: American Institute of Physics, 1969, 73 pp. Includes selected collections in electrical engineering.

0092 Yeudall, Bert, ed. **DIRECTORY OF TELEPHONE MUSEUMS.** Edmonton, Canada: Telephone Historical Centre, 1994 (7th ed.), 22 pp. Briefly annotated listing world-wide including contact people and telephone numbers.

2-A-4. Chronologies

0093 **A CHRONOLOGICAL HISTORY OF ELECTRICAL COMMUNICATION.** New York: National Electrical Manufacturer's Association, 1946, 160 pp. Some 106 pages of chronology plus appendix on specific firms. Index.

0094 Davis, Henry B. O. **ELECTRICAL AND ELECTRONIC TECHNOLOGIES: A CHRONOLOGY OF EVENTS AND INVENTORS.** Metuchen, NJ: Scarecrow Press, 1981–85 (3 vols.), 762 pp. Valuable reference through 1980, divided by decade, and within each section by subject matter. Bibliography, appendices, index.

See 0241. Dummer, **ELECTRONIC INVENTIONS AND DISCOVERIES.**

0095 **EVENTS IN TELECOMMUNICATIONS HISTORY.** Warren, NJ: AT&T Archives, 1992, 242 pp. Useful annotated chronology which emphasises Bell System developments concerning technology, business, regulation and service.

0096 Hudson, Robert V. **MASS MEDIA: A CHRONOLOGICAL ENCYCLOPEDIA OF TELEVISION, RADIO, MOTION PICTURES, MAGAZINES, NEWSPAPERS, AND BOOKS IN THE UNITED STATES.** New York: Garland, 1987, 435 pp. This media chronology also covers telecommunication. Use detailed subject index under AT&T, cable, cross media ownership, pay television, satellites, telecommunications, telephone, videotex, and other topics.

2-A-5. Biographical Sources

See 0012. **BIOGRAPHY AND GENEALOGY MASTER INDEX.**

See 0013. **BIOGRAPHY INDEX.**

0097 **DICTIONARY OF SCIENTIFIC BIOGRAPHY.** Ed. Charles Coulston Gillespie. New York: Charles Scribner's Sons, 1970–date. Vol. 18 published in 1990. Standard reference source including authoritative, signed scholarly entries, with bibliographies, on outstanding figures in science from earliest times to the present. Sponsored by the American Council of Learned Societies. Combined vol. 15–16 contains "Lists of Scientists by Field" ("Technology, Engineering") as well as comprehensive index: relevant entries under "Radio," "Telegraphy," "Telephony," "Television."

0098   Elliot, Clark A. **BIOGRAPHICAL INDEX TO AMERICAN SCIENCE: THE SEVEN-TEENTH CENTURY TO 1920.** New York: Greenwood, 1990, 300 pp. Cross references entries for science and technology figures included in over 100 major biographical dictionaries and collections, such as DICTIONARY OF AMERICAN BIOGRAPHY, NATIONAL CYCLOPEDIA OF AMERICAN BIOGRAPHY, WHO WAS WHO IN AMERICA, and BIOGRAPHY AND GENEALOGY MASTER INDEX. "Index of Names by Scientific Field" identifies relevant figures under "Engineering, Electrical," "Invention," "Telegraphy," and others.

0099   Ingham, John N. **BIOGRAPHICAL DICTIONARY OF AMERICAN BUSINESS LEADERS.** Westport, CT: Greenwood, 1983, 4 vols. Entries for 1,159 "historically most significant business leaders" (p. ix), with bibliographies. Several useful indexes: "American Business Leaders According to Industry" lists relevant entries under "Equipment Manufacturers—Electrical machinery," "—Typewriters and Business Machines,"   "—Electronics Manufacturers"; and "Public Utilities." Company index. See "Telegraph industry" and "Telephone industry" in General index.

2-B. Museums, Archives, and Libraries

See 2-A-3 for comprehensive directories

2-B-1. Museums

0100   **Alexander Graham Bell National Historic Park.** Chebutco St., Baddeck, Nova Scotia, B0E 1B0, Canada. Voice: 902 295 2069. Collections include daily records and artifacts related to Bell's experiments and personal memorabilia (1880s–1920s).

0101   **Antique Wireless Association, Inc. A.W.A. Electronic-Communication Museum.** Bloomfield, NY 14469. Mail address: Main St., Holcomb, NY 14469. Voice: 716 657 7489. Collections on radio communication and electronics; includes library and archives (access by appointment); publishes annual review, AWA REVIEW (1986–date); and quarterly journal, OLD TIMERS BULLETIN (1960–date).

0102   **Bell Homestead Museum.** 94 Tutela Heights Rd., Brantford, Ontario, N3T 1A1, Canada. Voice: 519 756 6220. Site of Bell family home and first telephone business office in Canada. Administered in part by Telephone Pioneers of America and Bell Canada.

0103   **Broadcast Pioneers Library.** Hornbake Library, University of Maryland College Park, MD 20742. Voice: 301 405 9160 or 301 405 9255. Papers of Elmo Neale Pickerill (MS 78–1182) includes materials on wireless telegraphy and radio communication related to Lee de Forest, Marconi, and others. Also relevant is papers of Federal Communications Bar Association (MS 79–1172). The collection's strengths are holdings related to radio and television broadcasting, programing, and personalities.

0104   **Edison Birthplace Museum.** Edison Birthplace Association, 9 N. Edison Dr., Milan, OH 44846. Voice: 419 499 2135. Edison family home includes Edison memorabilia and working models of his inventions.

0105   **Edison National Historic Site.** Main St. at Lakeside Ave., West Orange, NJ 07052. Voice: 201 736 5050. Fax: 201 736 8496. Established 1956; operated by National Park Service. Site was Edison's home and lab. Archives includes massive collection of Edison's papers from 1868–1931 (MS 66–797)—more than 500,000 items––containing personal and business correspondence, laboratory notebooks, record books, sketches, patents, and other documents related to the telephone, phonograph, and Edison's other inventions. Publishes many books on Edison and electrical inventions.

0106   **Franklin Institute Science Museum.** 20th and The Parkway, Philadelphia, PA 19103. Voice: 215 448 1200. Museum collections include original inventions of Franklin, Morse, Edison, and others. Franklin Institute Library's Historical Collections includes papers on David Galen McCaa (MS 63–156) [radio engineering]. A. Michael McMahon and Stephanie A. Morris's TECHNOLOGY IN INDUSTRIAL AMERICA: THE COMMITTEE ON SCIENCE AND THE ARTS OF THE

FRANKLIN INSTITUTE, 1824–1900 (Wilmington, DE: Scholarly Resources, 1977) indexes the 28-reel microfilm edition of the committee's records, identifying correspondence, minutes, and reports relating to telegraphy, telephony, and other electrical inventions.

0107   **French Cable Station Museum in Orleans.** 41 Rte. 28, Orleans, MA 02653. Mail address: P.O. Box 85, Orleans, MA 02653. Voice: 508 240 1735. Museum is Cape Cod site that terminated the cable laid from France to Eastham in 1879 and later cables; collections on submarine cables. Publishes descriptive booklet.

0108   **Historic Speedwell.** 333 Speedwell Ave., Morristown, NJ 07960. Voice: 201 540 0211. Fax: 201 540 0476. Museum is site of first electro-magnetic telegraph demonstration; collections include telegraphs, papers of Alfred Vail, portraits of Morse. Published Cam Cavanaugh's AT SPEEDWELL IN THE 19th CENTURY (1981) and video SPEEDWELL AND THE TELEGRAPH (1988) as well as newsletter, NEWS FROM HISTORIC SPEEDWELL.

0109   **Historical Electronics Museum, Inc.** 920 Elkridge Landing Rd., Linthicum, MD 21090. Mail address: P. O. Box 1693, MS 4610, Baltimore, MD 21203. Voice: 410 765 3803; 765 2345. Fax: 410 765 0240. Founded 1973; opened to the public 1983. Emphasis on military communications. Collections on advanced electronics, communications, radar, countermeasures, microwave, satellites; includes library and archives; publishes quarterly newsletter, REFELECTIONS (1991–date).

0110   **Motorola Museum of Electronics.** 1297 E. Algonquin Rd., Schaumburg, IL 60196. Voice: 708 576 6559. Fax: 708 576 6401. Founded 1986. Exhibits emphasize Motorola's role in evolution of electronic communication; facility includes library and corporate archives (see 0156).

0111   **Museum of Independent Telephony.** 412 S. Campbell, Abilene, KS 67410. Mail address: P. O. Box 625, Abilene, KS 64710. Voice: 913 263 2681. Library and archives; affiliated with United Telecommunications and Antique Telephone Collectors Association; collections (1875–date) on independent (non AT&T) telephone systems history, includes oral histories; publishes TALES OF TELEPHONY.

0112   **Museum of Science and Industry.** 57th St. and Lake Shore Dr., Chicago, IL 60611. Voice: 312 684 1414. Fax: 312 684 7141. A major permanent collection on telecommunications technology.

0113   **Museum of Transportation.** 3015 Barrett Station Rd., St. Louis, MO 63122, Voice: 314 965 7998. Founded 1944 (formerly National Museum of Transport); collections on communication devices, history of communication; includes library.

0114   **National Museum of Communications.** 6305 N. O'Connor, Ste. 123, Irving, TX 75039. Voice: 214 556 1234. Fax: 214 644 2473. Collections on telegraphs and telephones; includes library.

0115   **New England Wireless and Steam Museum.** Frenchtown Rd., East Greenwich, RI 02818. Mail address: 697 Tillinghast Rd., East Greenwich, RI 02818. Voice: 401 884 1710. Fax: 401 884 0683. Collections on electric communication equipment, wireless, telegraph, telephone, including Lloyd Espenshied, Thorn Mayes, Edward Raser and A. C. Goodnow collections of textbooks on wireless and telegraph; published Mayes' WIRELESS COMMUNICATIONS IN THE UNITED STATES: THE EARLY OPERATING COMPANIES (1989); includes library.

0116   **Signal Museum.** Fort Gordon, GA 30905. Voice: 404 791 3856. Maintained by U.S. Army Signal Center, formerly Signal Corps. Collections cover military communications from Civil War to modern era.

0117   **Telephone Pioneer Communications Museum of San Francisco.** New Montgomery St., Ste. 111, San Francisco, CA 94105. Mail address: 1515 19th Ave., San Francisco, CA 94122. Voice: 415 542 0182. Fax: 415 495 4133. Founded 1968; substantially refurbished and partially relocated in 1989. Affiliated with Pacific Bell, AT&T, and Telephone Pioneers of America; collections on telephones; includes library and archive (housed at 1515 Nineteenth Ave., San Francisco 94122; voice: 415 661 6469; Fax: 415 661 1077) which can be used by appointment. Publishes THE TRANS-MITTER (1989–date).

0118   **U.S. Army Communications-Electronics Museum.** Kaplan Hall, Bldg. 275, Fort Monmouth, NJ 07703. Voice: 908 532 2440. Museum site of Camp Alfred Vail, World War I Signal Corps training camp; formerly U.S. Army Signal Corps Museum; collections on telegraph and radio; library and archives holds papers on development of U.S. Army communications and electronics equipment, records of Telegraph Battalion (World War I), U.S. Veterans Signal Corps Association. Notable collections include papers of Albert James Myers (MS 67–145) [founder of Signal Corps]; and Signal Corps Veterans Association (MS 67–579).

2-B-2. National Libraries and Archives

0119   **Library of Congress.** Manuscript Division, 10 First St. SE, Washington, DC 20540. Voice: 202 707 5000. Fax: 202 287 5844. With the National Archives (0120), the Library of Congress holds the most extensive and important collections of documents on telecommunication in the U.S. Preeminent are the papers of Samuel Morse (MS 71–1385) and Alexander Graham Bell (MS 78–1688). A selection of other important collections includes the papers of Emile Berliner [telephone transmitter and microphones]; Albert Sidney Burleson (MS 60–160) [federal control of the telegraph as seen by a Postmaster General]; Lee de Forest (MS 59–25) [radio, Marconi]; Allen B. Dumont Laboratories (MS 68–2021) [television engineering, Federal Communications Commission]; Stanford Caldwell Hooper (MS 68–2043) ["Father of Naval Radio," military communication]; Frederic Eugene Ives and Herbert Eugene Ives (MS 67–612) [Bell Telephone Laboratories, wire telephoto services, color television]; Mahlon Loomis (MS 62–4526) [wireless telegraphy, Loomis Aerial Telegraph Company]; and Alfred Vail (MS 64–1582) [Morse and telegraphy].

0120   **National Archives.** Constitution Ave. and 8th St. NW, Washington, DC 20408. Voice: 202 501 5402. Fax: 202 523 4357. World-Wide Web: http://www.nara.gov. Contains the largest and most important collection of documents related to telecommunication. The most prominent record groups include the records of the Federal Communication Commission (RG 173) detailed in Albert W. Winthrop's PRELIMINARY INVENTORY OF THE RECORDS OF THE FEDERAL COMMUNICATIONS COMMISSION (1956); Commerce Department (RG 40), described in Forrest Robert Holdcamper's PRELIMINARY INVENTORY OF THE GENERAL RECORDS OF THE DEPARTMENT OF COMMERCE (1964); Signal Corps (RG 111), described in Mabel E. Deutrich's PRELIMINARY INVENTORY OF THE RECORDS OF THE OFFICE OF THE CHIEF SIGNAL OFFICER (1963); National Bureau of Standards (RG 167), described in William J. Lescure's PRELIMINARY INVENTORY OF THE RECORDS OF THE BUREAU OF STANDARDS (1964); and Rural Electrification Administration (RG 221) [rural telephones], described in Patrick G. Garabedian's PRELIMINARY INVENTORY OF THE RECORDS OF THE RURAL ELECTRIFICATION ADMINISTRATION (1977). Other record groups of interest include Office of Scientific Research and Development (RG 227); National Technical Information Service (RG 422); and Office of Technology Assessment (RG 444). Other National Archives collections are described in GUIDE TO THE NATIONAL ARCHIVES OF THE UNITED STATES (1974), which is updated in turn by GUIDE TO RECORDS IN THE NATIONAL ARCHIVES series of 11 regional guides (1989–90). Index to record groups (by name, office, and RG number) is LIST OF RECORD GROUPS IN THE NATIONAL ARCHIVES AND THE FEDERAL RECORDS CENTERS (1984). Specialized guides identifying Congressional documents related to telecommunication include Robert W. Coren's GUIDE TO THE RECORDS OF THE UNITED STATES SENATE IN THE NATIONAL ARCHIVES, 1789–1989 (Washington: U.S. Senate, 1989); and Charles E. Schamel's GUIDE TO THE RECORDS OF THE UNITED STATES HOUSE OF REPRESENTATIVES IN THE NATIONAL ARCHIVES, 1789–1989 (Washington: U.S. House of Representatives, 1989); as well as Brightbill's COMMUNICATIONS AND THE UNITED STATES CONGRESS (0378).

0121   **Dwight D. Eisenhower Library.** Abilene, KS 67410. Voice: 913 263 4751. Collections include papers of Thomas B. Larkin (MS 73–482) [World War II communications] and Donald A. Quarles (MS 76–1873) [Bell Telephone Laboratories, American Institute of Electrical Engineers], as well as collections on rural electrification.

0122   **Franklin D. Roosevelt Library.** 259 Albany Post Rd., Hyde Park, NY 12538. Voice: 914 229 8114. Collections with papers relevant to telecommunications include papers of Wayne Coy [FCC Chairman], John M. Carmody, and Morris Llewellyn Cooke [Rural Electrification Administration]; and Alexander Sachs [economics and finances of American Telephone and Telegraph Company, New York Telephone Company, RCA, and others].

0123   **Gerald R. Ford Library.** 1000 Beal Ave., Ann Arbor, MI 48109. Voice: 313 668 2218. Collections include papers of Frederick Lynn May [Office of Telecommunications Policy].

0124   **Harry S. Truman Library.** Independence, MO 64050. Voice: 816 833 1400. Includes papers of James Edwin Webb (MS 65–149) [National Aeronautics and Space Administration].

0125   **Herbert Hoover Presidental Library.** Parkside Dr., West Branch, IA 52358. Voice: 319 643 5301. Papers of Hoover (MS 70–185, MS 70–186) include materials on radio regulation in 1920s and formation of the Federal Radio Commission.

0126   **John F. Kennedy Library.** Columbia Point, Boston, MA 02125. Voice: (617) 929–4500. Collections include microfilmed documents of Federal Communication Commission, 1961–1963 [selected speeches of Newton Minow, FCC Chairman, files of Genral Counsel, Legislative Division, and President's reports]; and papers of Myer Feldman [telecommunications]; Lee C. White [Communications Satellite Corporation]; and oral history interviews of E. William Henry [FCC Chairman], Lee Loevinger [FCC Commissioner], Frederick R. Kappel [AT&T Chairman].

0127   **Smithsonian Institution.** National Museum of American History, Archives Center, 12th and Constitution Ave. NW, Washington, DC 20560. Voice: 202 357 3270. Fax: 202 787 2866. A small selection of important collections in history of electronics and telecommunications include Vail Telegraph Collection [Morse and telegraphy, Magnetic Telegraph Company]; Joseph Henry (MS 72–1239) [electromagnetism, Morse, and telegraphy]; Electrical Science and Technology Collection, Anglo-American Telegraph Company (MS 80–195) [Atlantic cables]; Western Union Telegraph Company (MS 80–209); Cyrus West Field (MS 80–313) [Atlantic cable]; Lloyd Espenschied (MS 80–200) [communications engineering]; Elisha Gray (MS 80–201) [Bell and Western Electric Company]; William Joseph Hammer (MS 80–202) [Vail, Edison, Bell, telephones]; William A. Dall (RU 7073) and Western Union Telegraph Expedition (MS 78–904) [Russian telegraph expedition, 1865–1868]; George H. Clark (MS 80–197) [National Electric Signaling Company, RCA, Fessenden]; Allen Balcom Dumont (MS 80–198) [radio and television standards and policy]; Harold R. Roess (MS 80–205) [radio navigation, antennas]; Kenneth M. Swezey (MS 80–207) [Tesla, radio industry]; and William Dandridge Terrell [Federal Communications Commission, Department of Commerce]. Several of these collections are described in GUIDE TO MANUSCRIPT COLLECTIONS IN THE NATIONAL MUSEUM OF HISTORY AND TECHNOLOGY (1978). Guides to specific collections include Robert S. Harding's REGISTER OF THE GEORGE H. CLARK RADIOANA COLLECTION, c. 1880–1950 (1985); and REGISTER OF THE WILLIAM J. HAMMER COLLECTION, c. 1874–1935, 1955–1957 (1986).

See 0191. **United States Air Force Academy.**

0128   **United States Military History Institute Library.** Carlisle Barracks, PA 17013-5008. Voice: 717 245 3611. Papers of the Signal Corps, 1860–1950, includes materials about the Corps from the Civil to Cold wars and beyond. Collections related to the Civil War Signal Corps include papers of George W. Callahan (MS 76–1252), John R. Mitchell (MS 79–1537), and Seibert Family (MS 79–1571). Papers of Bruno J. Rolak includes materials on heliograph communications during the Apache Wars. Papers of George A. Marshall includes materials on the telegraph and Signal Corps in Puerto Rico (1864–1898). Papers of William M. Chubb (MS 75–823) includes materials on civilian Signal Corps employment before and after World War I. Other collections related to World War I include papers of George O. Squier. Collections related to World War II Signal Corps also include Richard Bartholomew Moran (MS 79–1544) [World War II Signal Corps]. Collections related to later military communications include papers of Hans K. Ziegler [Signal Corps' Electronics Command and civilian electronics industry]; Hugh F. Foster [U.S. Army Electronics Command]; a photocopy of the Director of Defense Research and Engineering's "Program plan for development leading to the advanced defense communications satellite project (ADCSP)" (1964–1965); and Robert J. Lilley's "Survey of Cambodian communications" (1970) [United States Military Assistance Command, Vietnam (USMACV)].

2-B-3. State and Local Archives, Libraries, Historical Societies, etc.

0129   **Alaska Historical Society Library.** State Office Bldg., P.O. Box 11057, Juneau, AK 99811–0571. Voice: 907 465 2925. Fax: 907 465 2665. Includes records of the Alaska Communication System, 1900–1970 [telephone services of U.S. Army Signal Corps and Air Force Communications Service].

0130   **Arizona Historical Society Research Library.** 949 East Second St., Tucson, AZ 85719. Voice: 602 628 5774. Collections include papers of Bruce M. Stanley [letters and drawings on duplex telegraphy and telephony].

0131   **Atlanta Historical Society.** Society Library Archives, Atlanta, GA 30305. Voice: 404 261 1837. Fax: 404 238 0669. Incudes papers of Benjamin M. Bailey 1901–1945 (MS 79–51) [World War I Signal Corps]; Broughton W. Benning (MS 79–60) [American Radio Relay League]; and Southern Bell Telephone and Telegraph Company, 1909–1953 (MS 79–253).

0132   **Bridgeport Public Library.** Historical Collections, 925 Broad St., Bridgeport, CT 06604. Voice: 203 576 7777. Fax: 203 576 8255. Labor Records collections include papers of Louis J. Santoianni, 1945–1975, related to the United Electrical Workers, Local 203, and International Union of Electrical Workers, Local 203, General Electric Company.

0133   **California State Library.** 914 Capitol Mall, Library and Courts Bldg., P.O. Box 942837, Sacramento, CA 94237-0001. Voice: 916 654 0174. Fax: 916 654 0064. Includes papers of Joseph Mora Moss, 1800–1899 relating to California State Telegraph Company in San Francisco.

0134   **Chicago Historical Society.** Archives and Manuscripts, Clark St. at North Ave., Chicago. IL 60614. Voice: 312 642 4600. Collections include papers of Sterling Morton [printing teletype system]; and Harry R. Booth [American Telephone and Telegraph, Illinois Bell Telephone Company, Federal Communications Commission, telephone workers pensions]. Collections related to labor unions include papers of Ernest DeMaio, 1936–1974 [president of United Electrical, Radio, and Machine Workers (UE), District 11]; and papers of legal partners David Rothstein and Irving Meyers, ca. 1940–1960 [UE].

0135   **Colorado Historical Society.** Stephen H. Hart Library, 1300 Broadway, Denver, CO 80203. Voice: 303 866 2305. Fax: 303 866 5739. Includes papers of Leslie Morris Lytle (MS 76–41) [railroad telegraphers]; George E. Lawton [Western Union Telegraph Company]; Howard T. Vaille [Colorado Telephone Company, Mountain States Telephone and Telegraph Company]; Denver Pacific Railway and Telegraph Company, 1865–1868; Golden City and South Platte Railway and Telegraph Company, 1872–1880; and North Park and Wyoming Railroad and Telegraph Company, 1878.

0136   **Connecticut Historical Society.** 1 Elizabeth St., Hartford, CT 06105. Voice: 203 236 5621. Fax: 203 236 2664. Major collections includes papers of Samuel Colt (MS 66–406) [magnetic telegraph]; Hiram Percy Maxim (MS 78–1650 [American Radio Relay League]. Other collections include the papers of Milton Humphrey Bassett, 1859–1875 [Civil War military telegraph]; Morris Family, 1809–1946 [telephones in Syracuse, New York, 1879]; Gary Telephone Pay Station Company, 1921–1924.

0137   **Hagley Museum and Library.** Manuscripts and Archives, P.O. Box 3630, Greenville, DE 19807–0630. Voice: 302 658 2400. Fax: 302 658 0568. Collections include papers of Computer and Communications Industry Association IBM Antitrust Trials Records, 1969–1982; and White Dental Manufacturing Company, 1847–1970 [telegraph and telephone patents of Bell, Edison, Elisha Gray].

0138   **Massachusetts Historical Society Library.** 1154 Boylston St., Boston, MA 02215. Voice: 617 536 1608. Major collections include papers of Francis Blake, Jr. (MS 84–1991) [Bell Telephone Company] and William F. Channing [correspondence with Bell, development of telephone]. Other relevant collections include papers of John Albion Andrew Family, 1675–1906 [Western Union Telegraph Company]; Alexander Melville Clark, 1874–1875 [automatic chemical telegraphs]; and Abel Rathbone Corbin, 1727–1922 [transcontinental telegraph line construction].

0139   **Minnesota Historical Society.** Library and Archives, 160 John Ireland Blvd., St. Paul, MN 55102. Voice: 612 296 2143. Fax: 612 296 1004. Collections include records of Order of Railroad Telegraphers of North American, 1883–1956 (P9) [telegraphers trade unions].

0140   **New York Genealogical and Biographical Society Library.** 122 E. 58th St., New York, NY 10022–1939. Voice: 212 755 8532. Includes papers of Lloyd Espenshied (MS 76–644) [radio engineering, international communications].

0141   **New York Historical Society Library.** Manuscripts Department, 170 Central Park West, New York, NY 10024–5194. Voice: 212 873 3400. Fax: 212 874 8706. Collections include papers of Samuel Morse (MS 66–1812) [Morse's letters to his brother about telegraph]; and Henry O'Reilly (MS 61–1701) [erection of telegraph lines, dispute with Morse].

0142   **New York Public Library.** Rare Books and Manuscripts, Fifth Ave. and 42nd St., New York, NY 10018. Voice: 212 930 0800. Fax: 212 921 2546. Collections include papers of John Shaw Billings (MS 74–525) [correspondence with Bell]; Samuel H. Edwards (MS 70–1721) [Civil War telegraph operators]; Cyrus West Field (MS 76–1509) [correspondence with Morse and Joseph Henry]; Donald Monroe McNicol (MS 72–1052) [Civil War telegraphy, Old Time Telegraphers' and Historical Association].

0143   **North Carolina State Archives.** North Carolina Department of Cultural Resources, 109 E. Jones St., Raleigh, NC 27601. Voice: 919 733 2570. Fax: 919 733 8748. Collections include papers of Reginald Aubrey Fessenden (MS 66–1230) [Edison, Lee de Forest, A.E. Dolbear, wireless telegraphy, radio, television].

0144   **Ohio Historical Society.** Archives-Manuscripts Division, 1982 Velma Ave., Columbus, OH 43211–2497. Voice: 614 297 2510. Includes papers of Daniel H. Gard (MS 75–1003) [railroad telegraphy]; Pullan Family [Atlantic telegraph]; and Robert Henry Marriot [experimental radio technology and regulation]. Library holds labor union records, including papers of Communications Workers of America, Local 4371 (Marion, OH); International Union of Electrical, Radio, and Machine Workers, Local Local 746 (Columbus, OH), 1954–1979; and International Brotherhood of Electrical Workers, Local 54 (Columbus, OH), 1908–1914; Local 683 (Columbus, OH), 1924–1970; and Local 2020 (Reynoldsburg, OH), 1958–1971.

0145   **Pennsylvania Historical and Museum Commission.** Division of Archives and Manuscripts, Third and North Sts., P.O. Box 1026, Harrisburg, PA 17108-1026. Voice: 717 783 9898. Fax: 717 783 1073. Collections include papers of Warren J. Harder [Daniel Drawbaugh, telephone history]; Bell Telephone of Pennsylvania, 1886–1892 (MS 60–1787); and records of U.S. Circuit Court cases of American Bell Telephone, People's Telephone Company, Overland Telephone Company, Texas Pan Electric Telephone Company; and U.S. Patent Office Briefs of Bell and Francis Blake, 1880–1887.

0146   **Rochester Public Library.** Local History Division, 115 South Ave., Rochester, NY 14604-1896. Voice: 716 428 7300. Fax: 716 428 7313. Includes papers of Henry O'Reilly (MS 66–1915) [erection of telegraph lines, American Telegraph System, Western Union Telegraph Company].

0147   **State Historical Society of Wisconsin Library.** Archives Division, 816 State St., Madison, WI 53706. Voice: 608 264 6534. Fax: 608 264 6520. A major archive for telecommunications and mass communications research. A selection of most important collections include records of American Communication Association, 1934–1966, and ACA Locals 10 and 11 in New York, 1904–1973 [telegraphers trade unions, American Radio Telegraphists Association, Western Union Cable

Employees Association, Western Union Telegraph Company, RCA Communications, Congress of Industrial Organizations, Federal Communications Commission]; papers of Walter R. Baker (MS 68–2105) [National Television System Committee, Radio-Television Manufacturer's Association, color television]; Bornet L. Bobroff (MS 78–940) [Teleoptic Company]; Levi Burnell (MS 62–1803) [invention of telegraph]; Kenneth Allen Cox (MS 78–982) [Federal Communications Commission]; Malcolm Parker Hanson (MS 62–2357) [radio communication for Admiral Byrd's expeditions]; Herbert Clark Hoover (MS 68–2218) [President Hoover's role in radio regulation, Federal Radio Commission]; Lee Loevinger (MS 78–1009) [Federal Communications Commission]; James Lowth (MS 68–2259) [telephone receivers, Bell Telephone Company]; Frank Earl Mason (MS 65–1066) [short-wave radio, National Broadcasting Company]; Sig Mickelson (MS 78–1016) [communications satellites]; and National Broadcasting Company, 1930–1960 (MS 64–1619) [Engineering department records]. Collections described in Janice O'Connell's THE COLLECTIONS OF THE MASS COMMUNICATIONS HISTORY CENTER OF THE STATE HISTORICAL SOCIETY OF WISCONSIN AND THE COLLECTIONS OF THE WISCONSIN CENTER FOR FILM AND THEATER RESEARCH OF THE UNIVERSITY OF WISCONSIN (1979); SOURCES FOR MASS COMMUNICATIONS, FILM, AND THEATER RESEARCH: A GUIDE (1982); and F. Gerald Ham and Margaret Hedstrom's A GUIDE TO LABOR PAPERS IN THE STATE HISTORICAL LIBRARY OF WISCONSIN (1978).

0148   **Western Reserve Historical Society.** 10825 East Blvd., Cleveland, OH 44106–1788. Voice: 216 721 5722. Fax: 216 721 0645. Collections include records of Communications Workers of America, Local 4305, 1947–1975, and Local 4301, 1966–1970 [grievances against Ohio Bell Telephone Company]; William Andrew Manning [Western Union Telegraph Company, communications industry strikes and lockouts]; Joseph William Shiffman [Bell Electric Company, Ohio Bell Telephone Company, Ohio Federation of Telephone Workers, Telephone Construction Company, Cleveland, trade unions]; Jephta H. Wade (MS 75–1781) [submarine cables, Speed and Wade Telegraph Lines, Western Union Telegraph Company]; and William L. Gross (MS 62–4239) [United States Military Telegraph, Western Union Telegraph Company].

2-B-4. Corporate Archives

0149   **AT&T Archives.** AT&T Bell Laboratories. 600 Mountain Ave., Murray Hill, NJ 07974. Voice: 908 582 2003. Fax: 908 582 2975. Other archive locations: 5 Reinman Rd., Warren, NJ 07059–0647. Voice: 908 756 1586. Fax: 908 756 2105; and 295 North Maple Ave., Basking Ridge NJ 07920. Voice: 908 221 2816. Fax: 908 221 8174. Notable collections include Telephone Historical Collection; Western Electric Company, 1869–date; and papers of George C. Southworth [waveguides, microwaves]. Other collections include publications, records, and archives (company publications, press kits and press releases, product brochures, project filing cases and laboratory notebooks related to projects, historical book collection, executive biographies, speeches, personal papers, films and videos, 500,000 photographs, and historical artifacts and equipment. The Archives produces and publishes topical "Information Series" compilations and bibliographies. Offers fee-based customized research services. Access by written request.

0150   **Communications Satellite Corporation (COMSAT).** Library and Archives, 9560 Rock Spring Dr., Bethesda, MD 20817. Voice: 301 214 3682. Fax: 301 214 7128. Collections include records of communication satellite legislation; oral histories of COMSAT's founders; required financial filings with SEC; records of contract dealings; and COMSAT publications, brochures, and other public relations materials. Other historical files also held at COMSAT's Clarksburg technical laboratories. Access by request and appointment.

0151   **Dataproducts Corporation.** 6200 Conoga Ave., Woodland Hills, CA 91365. Voice: 213 888 4162; 709 1637. Records (1959–1980) for company and Univac, Telemeter Magnetics, Ampex Computers include publications, oral histories. Access by written request.

0152   **INTELSAT.** Library. 3400 International Dr. NW, Washington, DC 20008-3098. Voice: 202 944 6820. Library contains small archives of official publications, photographs, realia, and some oral histories. Not open to the public; access by request through Public Relations office.

0153   **ITT Corporation.** World Headquarters, Corporate Headquarters and Micrographics Department. 320 Park Ave., New York, NY 10022. Voice: 212 940 1690. Collections include Henri Busignies Papers, 1926–1960 [computer communications]; ITT Corporation Cable Collection, ca. 1890–1924 [records of calble-laying vessels of All-America Cable, Commercial Cable, Mackay Cable]; and ITT Corporation records, 1920–date. Access to collections by advanced written authorization.

0154   **MCI, Inc. Corporate Archive.** 1133 19th St. NW, Washington, DC 20036. Voice: 202 737 6290. Papers of key leaders and corporate records. Open to employees.

0155   **Microsoft.** Microsoft Corporate Archives, One Microsoft Way, Redmond, WA 98052. Voice: 206 936 8096. Tracks company history from 1975 to date as well as activities of Bill Gates and Paul Allen prior to formation of their partnership. Includes extensive collection of Microsoft and competitors' software, artifacts and ephemera, marketing and audiovisual material, and oral histories. Not currently open to the public; outside requests must be routed through Microsoft's Public Relations department (206 882 8080).

0156   **Motorola, Inc.** Corporate Offices, 1303 East Algonquin Rd., Schaumburg, IL 60196. Voice: 312 576 3495. Collections (1928–date) include publications, correspondence, advertising materials, product catalogs, oral histories; biographical materials on Paul Galvin, Robert Galvin, Daniel E. Noble, Elmer Wavering. Access by appointment.

0157   **Pacific Bell.** 1145 Larkin St., San Francisco, CA 94109. Voice: 415 661 6469. Collections (1896–date) include executive papers, publications, publicity materials. Part of Telephone Pioneers archives.

0158   **RCA.** David Sarnoff Research Center, 201 Washington Rd., Princeton, NJ 08543-5300. Voice: 609 734 2608. Now owned by SRI. David Sarnoff Library collections include materials on wireless technology and electronics.

0159   **Rockwell International.** Collins Commercial Telecommunications Division, 1200 North Alma Rd., Richardson, TX 75081. Voice: 214 996 5000. Engineering and business records (1950–date). Access by special permission.

0160   **Sprint International.** Christopher B. Newport Information Resource Center, 12490 Sunrise Valley Dr., Reston, VA 22096. Voice: 703 689 5388. Special collection of FCC tariff filings. Access by appointment.

0161   **Texas Instruments, Inc.** P.O. Box 655474, Mail Station 233, Dallas, TX 75265. Voice: 214 995 4458. Executive papers, business and technical/research records, publications. Access subject to approval.

0162   **UNISYS.** Information Research, 1 UNISYS Plaza, Detroit, MI 48232. Voice: 313 972 0318. Executive papers and records, publications, advertisements, product literature covering 1886–date.

2-B-5. College and University Archives

0163   **Boston University.** Special Collections, Mugar Memorial Library, 771 Commonwealth Ave., Boston, MA 02215. Voice: 617 353 3710. Fax: 617 353 2084. Historical manuscripts collection includes letters related to Bell's patents and the telephone.

0164   **Bowling Green State University.** Center for Archival Collections, 5th Floor, William T. Jerome Library, Bowling Green, OH 43403-0170. Voice: 419 372 2856. Fax: 419 372 7996. Collections include records of Communications Workers of America Local 4323, Tiffin, OH, 1953–1983; and the International Brotherhood of Electrical Workers, Local 986, Norwalk, OH, 1943–1979.

0165   **Case Western Reserve University Libraries.** Archive of Contemporary Science and Technology, 10900 Euclid Ave., Cleveland, OH 44106-7151. Voice: 216 368 2992. Fax: 216 368 6950. Collections include papers of Donald J. Angus (MS 68–903) [electrical recording devices]; William D. Buckingham (MS 71–1505) [Western Union]; William L. Gross (MS 62–4239) [Civil War military telegraphy, Western Union].

0166   **Catholic University of America.** Archives and Manuscripts, John K. Mullen Denver Memorial Library, 620 Michigan Ave. NE, Washington, DC 20064. Voice: 202 635 5065. Fax: 202 319 6101. Contains the Joseph Keenan Collection related to the International Brotherhood of Electrical Workers.

0167   **Columbia University.** Rare Books and Manuscripts, Butler Library, 535 W. 114th St., New York, NY 10027. Voice: 212 854 2231. The most important collections include papers of Edwin Howard Armstrong (MS 80–1888) [some 200,000 items related to research and inventions, frequency modulation, U.S. Air Force communications, Radio Club of America, RCA, Zenith Radio Corp., lawsuits, etc.]; Thomas Edison (MS 80–1900) [telegraph and telephone companies, Morse, Gold and Stock Telegraph Company]; and Samuel Morse (MS 66–1595) [Morse Centennial Celebration]. Other collections include papers of Eric Barnouw (MS 77–73) [communications history]; Beril Edelman (MS 80–1899) [defense and industrial electronics]; James Lawrence Fly [Federal Communications Commission, Board of War Communications]; Michael Idvorsky Pupin [research in telephony]; Nicola Tesla (MS 64–1364) [wireless transmitters]; and Egbert Harold Van Delden [correspondence with Lee de Forest]. The collection also includes oral histories by de Forest and others.

0168   **Cornell University.**
          1 Manuscripts and University Archives, John M. Olin Library, Ithaca, NY 14853-5301. Voice: 607 255 3530. Fax: 607 255 9346. Includes papers of Alonzo Cornell (MS 62–2343) [Albany and Buffalo Telegraph Company, Erie and Michigan Telegraph Company]; Ezra Cornell (MS 62–2374, MS 77–157, MS 77–158) [Morse telegraph line, New York and Erie Telegraph, Western Union Telegraph Company]; George Lucien Swift (MS 62–2197) [electromagnetism and telegraphy]; and Robert Henry Thurston (MS 62–4323) [Bell, Edison]. Additionally, numerous other collections cover electrical engineering research and education at Cornell.
          2 Labor-Management Documentation Center, Martin P. Catherwood Library, 144 Ives Hall, Ithaca, NY 14853-3901. Voice: 607 255 3183. Fax: 607 255 9641. Includes the records of the Telecommunications International Union, 1949–1986.

0169   **Dartmouth College.** Special Collections and Manuscripts, Baker Library, Hanover, NH 03755. Voice: 603 646 2236. Fax: 603 646 3702. Includes papers of Ernest Galen Andrews [Bell Telephone Laboratories]; Edward Kimball Hall, 1909–1932 (MS 76–1835) [American Telephone and Telegraph Company]; Gordon Ferrie Hull (MS 71–1016) [Bell Telephone Laboratories, radio development]; and George Robert Stiblitz (MS 76–1844) [Bell Telephone Laboratories, computer communications research].

0170   **Duke University.** Special Collections, William R. Perkins Library, Durham, NC 27706. Voice: 919 684 2034. Fax: 919 684 2855. Major collections include papers of Horatio Hubbell [transatlantic telegraph cable, Morse letters]; and James Carmichael Smith (MS 71–86) [submarine cables]. Other relevant collections are papers of James Martin Bell, 1799–1870 [Pennsylvania telegraph companies]; Coleman Family, 1806–1921 [telegraphs in Arkansas]; Snow Family, 1861–1865 [Civil War telegraph operators]; and Thomas Jordan, 1861–1885 [telegraphs in South Carolina].

0171   **Emory University.** Special Collections, Robert W. Woodruff Library, 540 Asbury Circle, Atlanta, GA 30322. Voice: 404 727 6861. Collections include papers of Samuel Morse (MS 77–246) [correspondence of Morse and his first wife].

0172   **George Washington University.** Special Collections Department, Melvin Gelman Library, 2130 H Street, NW, Washington, DC 20052. Voice: 202 994 7549. Fax: 202 994 1340. Collections include the papers of Joseph V. Charyk 1962–1985 [COMSAT]; papers of Peter H. Sera, 1964–1989

[NL Industries and SBS]; papers of Albert F. Murray, 1921–1984 [Hammond Radio Research Lab, submerged radio-controlled torpedo development].

0173   **Georgetown University.** Special Collections, Joseph Mark Lauinger Library, 1421 37th St. NW, Washington, DC 20057-1006. Voice: 202 687 7425. Fax: 202 687 7501. Collections include papers of Rogers Family [American Radio Relay League, American Telephone and Telegraph, Alexander Graham Bell, Emile Berliner, Rogers Telephone Company, United States Postal Printing Telegraph Company, Wanamaker Telephone Company]; and Otto E. Guthe [remote sensing, satellites].

0174   **Georgia State University.** Special Collections, William Russell Pullen Library, 103 Decatur St., Atlanta, GA 30303. Voice: 404 651 2185. Fax: 404 651 2508. The Southern Labor Archives collections include records of the Order of Railway Telegraphers; and Communications Workers of America Local 3250, 1935–1945 [Federation of Long Lines Telephone Workers] and District 3, 1953–1969 [grievance cases filed against Southern Bell Telephone and Telegraph Company]; the International Union of Electronic Workers; and Adair, Goldthwaite, Stanford, and Daniel law firm, 1955–1956 (MS 77–259) [Communication Workers of America, Southern Bell Telephone and Telegraph Company, trade unions].

0175   **Harvard University.**
        1 Manuscripts and Archives Department, Baker Library, Graduate School of Business Administration. Soldiers Field Rd., Boston, MA 02163. Voice: 617 495 6405. Fax: 617 495 6001. Collections include legal papers of Bell Telephone Company vs. Peter A. Dowd, 1879; and Western Union vs. American Bell, 1881–1892 (MS 60–168) [Western Union Telegraph Company U.S. Circuit Court case about invention of telephone]; and Western Electric Company, 1924–1934 (MS 78–258) [industrial and employee relations at Hawthorne Works in Chicago].
        2 Rare Books and Manuscripts, Houghton Library, Cambridge, MA 02138. Voice: 617 495 2441. Fax: 617 495 1376. Collections include papers of Samuel Batchelder [research on telegraphy and submarine cables]; and George Washington Pierce [wireless telegraphy]
        3 Harvard University Archives, Pusey Library, Cambridge, MA 02138. Voice: 617 495 2461. Relevant collections include the papers of Frederick V. Hunt (MS 76–1971) [communications engineering, Institute of Radio Engineers, Bell Telephone Laboratories, General Electric Company, Western Electric Company, RCA, Sylvania, and others]; and records groups for Harvard University Department of Physics, 1886–date [Radio Research Laboratory, U.S. Naval Radio School]; Howard H. Aiken Computation Laboratory, 1944–1961 [Bell Telephone Laboratories]; Harvard University Program on Information Resources Policy, ca. 1960–1980 [formerly Program on Information Technology and Public Policy]. Other Archives' collections include faculty papers related to electrical engineering research and education.

0176   **Johns Hopkins University Library.** Ferdinand Hamburger, Jr., Archives, Baltimore, MD 21218. Voice: 410 516 8325. Fax: 410 516 8596. Records of the President's Office includes correspondence with Bell. Other collections include papers of Henry Augustus Rowland [Rowlands Multiplex Printing Telegraph, Rowlands Telegraphic Company]; Francis White [International Telephone and Telegraph Corporation]; and Hugh L. Dryden [National Aeronautics and Space Administration].

0177   **Massachusetts Institute of Technology Libraries.** Institute Archives and Special Collections, Cambridge, MA 02139. Voice: 617 253 5136. Fax: 617 253 1690. Collections include papers of Vannevar Bush (MS 84–2072) [American Telephone and Telegraph]; Gordon Stanley Brown [television and computer engineering research]; James Rhyne Killian, Jr. [American Telephone and Telegraph, Corporation for Public Broadcasting]; Blatchford, Seward, and Griswold (MS 84–2070) [telegraph and telegraph cable patents]; Francis Bitter (MS 84–2069) [Bell Telephone Laboratory]; Max F. Millikan [international telecommunication cooperation]; MIT Industrial Liaison Program, 1948–1986 [Bell Telephone Laboratory, International Business Machines Corporation, International Telephone and Telegraph Corporation]; Walter Alter Rosenblith [electronics in biomedical engineering]; Norbert Wiener [Bell Telephone Laboratories, Institute of Radio Engineers, Vannevar Bush]; Homer Walter Dudley [Bell Telephone Laboratories research on speech transmission by wire, cable, and radio telephony]; and Ithiel De Sola Pool [communications research and policy]. Materials on electrical engineering research and education are found thoughout numerous record groups for MIT schools, departments, and laboratories.

0178   **New York University.** Tamiment Institute-Ben Josephson Library and Robert F. Wagner Labor Archives, Elmer Holmes Bobst Library, 70 Washington Square South, New York, NY 10012-1091. Voice: 212 998 2440. Includes papers of telephone workers at AT&T Longlines, Communications Workers of America, Local 1150 (1940–1978).

0179   **Oberlin College Library.** Special Collections, Oberlin, OH 44074. Voice: 216 775 8285. Fax: 216 775 8739. Most prominent collection is papers of Elisha Gray, 1873–1964 (part of the Oberlin College Physics Department Records), including correspondence with Bell and materials related to controversy over telephone patent. Collections of papers of Lloyd William Taylor, 1905–1980, also include materials about Gray and the telephone.

0180   **Pennsylvania State University.** Historical Collections and Labor Archives, Fred Lewis Pattee Library, University Park, PA 16802. Voice: 814 865 0401. Fax: 814 865 3665. Includes papers of Harry Block of the United Electrical Workers and the International Union of Electrical Workers; and oral histories related to the United Electrical Workers.

0181   **Radcliffe College.** Arthur and Elizabeth Schlesinger Library on the History of Women in America, 3 James St., Cambridge, MA 02138–3766. Voice: 617 495 8647. Collections include papers of Louise Barbour (MS 77–1725) [Army Signal Corps, telecommunication cables, telephone operators]; Chloe Owings (Pasadena Institute for Radio); Frieda Hennock Simons [Federal Communications Commission—first woman in major regulatory agency]; Sally L. Hackey [Bell Telephone Laboratory, National Organization for Women: American Telephone and Telegraph Task Force]. The Library's Black Women's Oral History Project collections include interview by Mae Graves Eberhardt, electronics worker and civil rights director of International Union of Electrical, Radio, and Machine Workers (IUE), District 3.

0182   **Rensselaer Polytechnic Institute.** Rensselaer Polytechnic Institute Archives, Folsom Library, 110 Eighth St., Troy, NY 12180–3590. Voice: 518 276 8310. Fax: 518 276 8559. Record group for RPI's Electrical Engineering Department, 1938–1979, includes materials on the Dumont Laboratories. Other record groups include materials for electrical engineering research and education.

0183   **Rutgers University.** Special Collections and Archives, Alexander Library, 169 College Ave., New Brunswick, NJ 08903. Voice: 908 932 7505. Fax: 908 932 7637. Major collection is the papers of International Union of Electrical, Radio, and Machine Workers from 1930s through the present. The Library serves as official repository for the IUE's archives. Collection also includes transcripts and exhibits from 1947 NJ Statuatory Board of Arbitration dispute of New Jersey Bell Telephone Company and the Telephone Workers Union of New Jersey; and papers of the Lockwood Long Distance Telephone and Telegraph Company of America, 1897–1913.

0184   **San Francisco State University.** Labor Archives and Research Center, J. Paul Leonard Library, 1630 Halloway Ave., San Francisco, CA 94132–1789. Voice: 415 469 1681. Fax: 415 338 6199. Includes records of the United Electrical Radio, and Machine Workers (UE) locals in East Bay, San Jose, and Sunnyvale (1944–1981).

0185   **Sangamon State University.** Special Collections, Norris L. Brookens Library, Springfield, IL 62794–9243. Voice: 217 786 6597. Fax: 217 786 6208. The Library's Telephone: Oral History Collection (MS 76–2095) includes transcripts of interviews by ten former employees of Illinois Bell Telephone Company and other telephone companies.

0186   **Smith College.** Special Collections, William Allan Neilson Library, Northapmton, MA 01063. Voice: 413 585 2910. Includes papers of Lorena Estelle Hermance, 1942–1985 related to U.S. Army Women's Auxiliary Corps and Signal Corps in World War II.

0187   **Stanford University.**
1 Hoover Institution on War, Revolution, and Peace, Stanford, CA 94305-6004. Voice: 415 723 2058. Fax: 415 723 1687. Collections include papers of 317th Field Signal Battalion, 1917–1972, and of George E. Stone related to World War I Signal Corps; George W. Gilman [international radio-telephone communications with Japan previous to WW II]; Norman Lee Baldwin, James A. Code, Jr., and Ernest A. Ewers (MS 75–621) related to World War II Signal Corps; Carll Herman Sturies [Korean War Signal Corps]; and Katherine Drew Hallgarten [international communications and space law].
2 Special Collections, Cecil H. Green Library, Stanford, CA 94305-6004. Voice: 415 723 9108. Fax: 415 725 4902. Collections include papers of William Webster Hansen (MS 67–2099, MS 76–1624) [research on microwaves, radar, Institute of Radio Engineers]; James Arthur Miller [Federal Telegraph Company, radio, sound recording, correspondence with Edwin H. Armstrong]; Frederick Emmons Terman [Institute of Radio Engineers, Harvard Radio Research Laboratory, microwaves]; and Cyril Frank Ewell (MS 64–726) [Poulsen Wireless Telephone and Telegraph Company, Lee de Forest, Hewlett-Packard]. Other record groups include materials on electrical engineering research and education at Stanford.

0188   **State University of New York at Albany Library.** Special Collections and Archives, 1400 Washington Ave., Albany, NY 12222-0001. Voice: 518 442 3600. Fax: 518 442 3576. Important collections related to labor unions include the records of International Union of Electronic, Electrical, Salaried, Machine, and Furniture Workers (IUE), Local 301, 1949–1989 [represents General Electric's first plant, in Schenectady]; and IUE Local 379, 1944–1987.

0189   **State University of New York at Stony Brook.** Special Collections, Frank Melville, Jr., Memorial Library, Stony Brook, NY 11794-3300. Voice: 516 632 7100. Fax: 516 632 7116. Charles Weston Hansell Papers, 1928–1967 (MS 71–1814) includes materials on Hansell's radio and television research and RCA's Radio Transmission Laboratory.

0190   **Union College.** Schaffer Library, Special Collections, Schenactady, NY 12308. Voice: 518 370 6277. Fax: 518 370 6619. Includes papers of Ernest Fredrik Werner Alexanderson (MS 76–1639, MS 80–338) [radio, television inventions, RCA]. Other collections include materials on electrical engineering research at Union College.

0191   **United States Air Force Academy.** Special Collections, Academy Library, Colorado Springs, CO 80840-5701. Voice: 719 472 2590. Fax: 719 472 4754. Collections include papers of George Owen Squier [Army Signal Corps, Trans-Pacific cable].

0192   **University of Alaska.** Rare Books, Archives, and Manuscripts, E.E. Rasmuson Library, Fairbanks, AK 99775-1000. Voice: 907 474 7224. Fax: 907 474 6841. Includes papers of Charles Stewart Farnsworth (MS 68–2) [Alaska telegraph lines].

0193   **University of Arkansas Libraries.** Special Collections, Fayetteville, AR 72701-1201. Voice: 501 575 4101. Fax: 501 575 5558. Collections include papers of Oren Harris [Congressional files related to communications, Allied Telephone Company].

0194   **University of California at Berkeley.** Bancroft Library, Berkeley, CA 94720. Voice: 510 642 3773. Fax: 510 643 7891. One of the major collections of materials for U.S. telecommunications history. Includes two major collections: Alaska Collection, ca. 1750–1957 (MS 65–1780), includes papers of George Chismore, Edward Conway, United States Coast Guard, and Ferdinand Westdahl related to the Western Union Telegraph Company's expedition for the Russian extension. Additionally, the papers of Charles Melville Scammon (MS 65–1171) are also related to the expedition. The other major collection is the History of Science Collection, including papers of Lewis M. Clement [radio]; Gerhard R. Fisher [radio engineering]; Leonard Franklin Fuller [Federal Telegraph Company, de Forest-Armstrong patents]; Charles Vincent Litton (MS 80–2281) [Federal Telephone and Radio Corporation]; Haraden Pratt [radio engineering, Federal Telegraph Company]; Samuel Silver [microwave electronics, space sciences, International Union of Radio Science]; Emil Jacob Simon [radio engineering, Lee de Forest]; Albert Vasseur [history of wireless telegraphy]; and Communications Satellite Corporation (COMSAT), 1963–1979 [domestic communication satellite systems]. Other collections include papers of John Oliver Ashton [early radio technology]; Wyoming Collection (MS 65–1794) [Pacific Telegraph Company]; Loring Dumas Beckwith [International Telephone and Telegraph Corporation]; Robert Dollar (MS 75–283) [Globe Wireless, Ltd.]; Benjamin G. Griffith (MS 80–2262) [radio operator]; Donald Knudsen Lippincott [radio and television patent law]; Herman Potts Miller, Jr. [Federal Telegraph Company, ITT]; Alexander Mathew Poniatoff [Ampex Company, magnetic tape recording]; and Vrian Associates, 1948–1972 [microwave electronics]. Collections related to western U.S. companies include papers of Deseret Telegraph Company, 1880 [Salt Lake City, Utah] and Mount Diablo Telephone Company, Clayton, California, 1912–1945. Several other collections of papers are related to Pacific Telephone and Telegraph Company, 1865–1866, 1920–1931.

0195   **University of California at Los Angeles.** Special Collections, University Research Library, 405 Hilgard Ave., Los Angeles, CA 90024-1575. Voice: 310 825 1201. Collections include papers of Walter Beyer (MS 73–60) [television engineering]. Oral History Program collection includes an interview with William Jarvis Carr, 1959 [Pacific Telephone and Telegraph Company].

0196   **University of Colorado Libraries.** Western Historical Collections, Box 184, Boulder, CO 80309-0184. Voice: 303 492 7511. Fax: 303 492 2185. Includes collections related to the International Typographical Union (1889–1986) which merged with the Communications Workers of America in 1987.

0197   **University of Connecticut at Storrs Library.** Historical Manuscripts and Archives, Storrs, CT 06269. Voice: 203 486 2219. Fax: 203 486 3593. The Connecticut Labor Archives includes papers of Nicholas J. Tomassetti [international representative of the United Electrical, Radio, and Machine Workers Union]; the 1942 charter of the United Telegraph Workers International Local 47 (East Hartford); and papers of members of the International Typographical Union locals.

0198   **University of Georgia.** Special Collections, Richard B. Russell Memorial Library, Athens, GA 30602. Voice: 404 542 0621. Fax: 404 542 6522. Collections include papers of Jasper Dorsey [Southern Bell Telephone Company]; also special collection of papers and records of Senator Richard B. Russell includes some telecommunication material.

0199   **University of Illinois at Urbana-Champaign Library.** Special Collections, 1408 W. Gregory Dr., Urbana, IL 61801. Voice: 217 333 0790. Fax: 217 244 0398. Collections includes papers of Harry Applewhite [satellite communications]; John Bardeen (MS 72–1500) [Bell Telephone Laboratories, transistors, semiconductor research]; Sidney A. Bowhill (MS 77–1166) [radio and satellite engineering]; Earl Mulford Hughes [Illinois Bell Telephone Company]; Joseph T. Tykociner (MS 71–1170) [wireless telegraphy and electronics]; George W. Swenson [space radio research, satellites]; William J. Wardall (MS 75–697) [Associated Telephone Utilities Company, United Telephone and Electric Company]; and Arthur Gordon Webster [telegraphy]. Several other collections include papers related to space communications and electrical engineering education and research.

0200   **University of Iowa Library.** Special Collections, Iowa City, IA 52242. Voice: 319 335 5876. The major collection is the papers of the Communications Workers of America Oral History project, which includes transcripts of 89 interviews with union officials from 1917–1961 (conducted in 1969–1972) extending 220 hours, with a 2,000 page subject index.

0201   **University of Kansas.** Kenneth Spencer Research Library, Kansas Collection, Lawrence, KS 66045-2800. Voice: 913 864 4274. Papers of Kansas District Court (29th District), 1858–1889 (MS 80–375) includes lawsuits related to Great Western Telegraph Company.

0202   **University of Massachusetts at Amherst Library.** Archives and Manuscripts, Amherst, MA 01003. Voice: 413 545 2780. Contains several important collections related to the electrical labor unions, including the records of International Union of Electrical, Radio, and Machine Workers (IUE) Local 206, 1942–1986; and IUE, Local 278, 1942–1983. Also holds papers of Sidney Topol [microwave, satellite, and cable technologies at Raytheon and Scientific Atlanta].

0203   **University of Michigan, Ann Arbor.** Bentley Historical Library, 1150 Beal Ave., Ann Arbor, MI 48109–2113. Voice: 313 764 3482. Fax: 313 763 5080. Collections include papers of James Graig

Watson (MS 65–627) [invention of telephone, dispute between Bell and Gray]; George H. Harmon [Bell Telephone Laboratory]; Helen W. Berthelot (MS 80–977) [communication satellites, Communication Workers of America]; George Owen Squier (MS 65–567) [U.S. Signal Corps, transoceanic cables]; Thomas George Long (MS 80–1372) [telephone rates court cases]; Delano Family [National Bell Telephone Company]; William O. Gassett (MS 80–1206) [aviation radio communication]; and Arthur Pound (MS 65–496, MS 80–1528) [RCA]. Several other collections include materials on Michigan telephone companies. Guides include John D. Stevens' BIBLIOGRAPHY OF UNPUBLISHED RESOURCES ON THE HISTORY OF MASS COMMUNICATION IN THE MICHIGAN HISTORICAL COLLECTIONS (1975).

0204   **University of Minnesota, Twin Cities Libraries.** Charles Babbage Institute, Walter Library, 117 Pleasant St. SE, Minneapolis, MN 55455. Voice: 612 624 5050. Collections include papers of John L. Hill [teletype message systems]; Robert M. Kalb [research at Bell Telephone Laboratories and elsewhere on telecommunications transmission and terminal equipment for communications systems]; Computer and Communications Industry Association Anti-Trust Records, ca. 1957–1982; and Auerbach Associates, Inc., Records, 1958–1976 [consulting reports, planning studies, etc., for AT&T, ITT, RCA, U.S. Army Strategic Communications Command]. Other materials descibed in Bruce Bruemmer's GUIDE TO THE ORAL HISTORY COLLECTION OF THE CHARLES BABBAGE INSTITUTE (1986).

0205   **University of North Carolina at Chapel Hill.** Manuscripts Department and Southern Historical Collections, Wilson Library, Chapel Hill, NC 27599-3930. Voice: 919 962 1301. Fax: 919 962 0484. Collections include papers of Matthew Harrison [cipher and telegraph codes]; Edward McCrady L'Engle (MS 64–1047) [International Ocean Telegraph Company]; Paul Newman Guthrie [telecommunication trade unions]; Goodrich Wilson Marrow (MS 61–3652) [North Carolina telegraph and telephone companies]; and Archibald Lee Manning Wiggins [American Telephone and Telegraph Company, 1948–1952].

0206   **University of North Dakota.** Special Collections, Chester Fritz Library, Grand Forks, ND 58201-0175. Voice: 701 777 2617. Fax: 701 777 3319. Includes papers of Edwin Lee White (MS 70–376) [FCC, radio communication]; and William Langer [rural telephone legislation in U.S. Senate].

0207   **University of Notre Dame.** Special Collections, Hesburgh Library, Notre Dame, IN 46556. Voice: 219 239 5252. Fax: 219 239 6772. Includes papers of Rudolph Otto Probst, 1900–1918 [telephony].

0208   **University of Oregon Library.** Special Collections, Eugene, OR 97403-1299. Voice: 503 346 3053. Fax: 503 686 3094. Collections include papers of Leonard A. Andrus (MS 76–1278) [American Telephone and Telegraph Company, Bell Telephone System]; Baker Family (MS 64–685) [rural telephone lines]; Dale M. Newton (MS 76–1318) [World War I Signal Corps]; and Claude R. Lester (MS 70–1797) [valuation and ratemaking studies of Pacific Telephone and Telegraph Company].

0209   **University of Pittsburgh.** Archives of Industrial Society, Hillman Library, Pittsburgh, PA 15260. Voice: 412 648 7710. Fax: 412 648 7891. Official repository for the records of the United Electrical, Radio, and Machine Workers. Record groups in this collection include the UE District/ Local Series, 1935–1965; UE Publications Series, 1935–1965; UE Organizers Series, 1935–1960; and UE Conference Board Series, 1935–1965. Other notable specific major collections related to United Electrical, Radio, and Machine Workers, Local 601, East Pittsburgh include the papers of Margaret Darin Stasik (MS 77–2044), Thomas J. Quinn (MS 77–2043); and United Electrical, Radio, and Machine Workers of America, Local 601, 1949–1952 (MS 70–1846).

0210   **University of Rochester.** Special Collections, Rush Rhees Library, Rochester, NY 14627. Voice: 716 275 4461. Fax: 716 473 1906. Collections include papers of Hiram Sibley [Western Union Telegraph Company, Russian-American Telegraph Company]; William Roy Vallance (MS 69–1766) [telegraph law and legislation]; and Freeman Clarke (MS 84–1054) [Western Union Telegraph Company].

0211   **University of Southern Indiana Library.** Special Collections and University Archives, 8600 University Blvd., Evansville, IN 47712-3595. Voice: 812 464 1824. Fax: 812 465 1693. Collections include papers of United Electrical, Radio, and Machine Workers of America, 1949–1954 (MS 78–756).

0212   **University of Texas at Arlington Library.** Special Collections, Arlington, TX 76019. Voice: 817 273 3000. Fax: 817 273 3392. Papers in the Texas Labor Archives collections include records of International Brotherhood of Electrical Workers, Local 66 (Houston, TX), 1909–1962 (MS 74–1059); Local 69 (Dallas, TX), 1946–1971 (MS 73–880); Local 72 (Waco, TX), 1912–1946 (MS 72–654); Local 156 (Fort Worth, TX), 1919–1969 (MS 73–881); Local 278 (Corpus Christi, TX), 1939–1954 (MS 72–1325); and Local 850 (Lubbock, TX), 1929–1966 (MS 72–1326). Other collections related to Communications Workers of America, District 6, Southwestern Bell, Telephone Pioneers Association, and other topics include Blanche Wells (MS 77–2063) and Robert W. Staley. Photograph Collection also includes materials on Communications Workers of America depicting events in Texas, 1938–1966.

0213   **University of Texas at Austin.** Eugene C. Baker Texas History Center, Sid Richardson Hall, Austin, TX 78713-7330. Voice: 512 495 4515. Fax: 512 495 4542. Collections include papers of Albert Sidney Burleson (MS 69–1956) [International Wire Communication Conference, 1920; United States Telegraph and Telephone Administration].

0214   **University of Utah.** Manuscripts Division, Marriott Library, Salt Lake City, UT 84112. Voice: 801 581 8558. Includes papers of Philo Farnsworth [invention of television] and other materials on electronic media in Utah and intermountain region. Guide is Karin Hardy's GUIDE TO THE BROADCAST COLLECTIONS (1993).

0215   **University of Virginia.** Special Collections, Alderman Library, Charlottesville, VA 22903-2498. Voice: 804 924 3026. Fax: 804 924 4337. Collections include Gaines Family Papers, 1841–ca. 1890 [Transatlantic cable].

0216   **University of Washington Library.** Manuscripts and University Archives, Seattle, WA 98195. Voice: 206 543 1879. Papers of Charles S. Hubbell (MS 65–1037) include materials related to Western Union Russian Telegraph Expedition.

0217   **University of Wyoming Library.** American Heritage Center, Laramie, WY 82071-3334. Voice: 307 766 3279. Fax: 307 766 3062. Collections include papers of Eldridge Reeves Johnson (MS 77–1269) [Victor Talking Machine Company]; and records of the Medicine Bow Telephone Company, 1906–1913.

0218   **Virginia Polytechnic Institute and State University.** Special Collections, Carol M. Newman Library, Blacksburg, VA 24061–9001. Voice: 703 231 5593. Fax: 703 231 9263. Collections include papers of W. Graham Claytor [ham radio]; records of the Cincinnati and Eastern Telegraph Company, 1877–1902 (part of Norfolk and Western Railway Archives); and papers of Marjorie Rhodes Townsend, 1966–1980 (part of the Archives of American Aerospace Exploration) [communications satellites].

0219   **Washington State University Library.** Manuscripts, Archives, and Special Collections, Pullman, WA 99163-5610. Voice: 509 335 4557. Fax: 509 335 0934. Papers of Eugene Willis Greenfield [wire and cable insulation]; and Homer Jackson Dana (MS 71–1979) [communications research, facsimile transmission].

0220   **Wayne State University Library.** Archives of Labor History and Urban Affairs, Detroit, MI 48202. Voice: 313 577 4024. Fax: 313 577 5525. Papers of the Michigan AFL-CIO, 1939–1958 (MS 72–837) relate to expulsion of United Electrical, Radio, and Machine Workers of America from AFL-CIO. Collections also include the records of the Michigan offices of the Communications Workers of America, District 4; and papers of Helen W. Berthelot [Communications Workers of America].

0221   **Wells College.** Special Collections, Louis Jefferson Long Library, Aurora, NY 13026-0500. Voice: 315 364 3351. Fax: 315 364 3412. Papers of Edwin Barber Morgan (MS 80–2382) includes materials on establishment of express and telegraph companies.

0222   **West Virginia University Library.** West Virginia and Regional History Collection, Colson Hall, Morgantown, WV 26506. Voice: 304 293 3536. Fax: 304 393 3303. Collections include papers of Allen D. Frankenberry (MS 77–679) [Civil War Signal Corps]; and Chesapeake and Potomac Telephone Company, 1936–1938 (MS 59–174).

0223   **Wright State University Library.** Special Collections, Colonel Glenn Hwy., Dayton, OH 45435. Voice: 513 873 2380. Fax: 513 873 2526. Includes records of Communications Workers of America, Local 4322, 1936–1975.

0224   **Yale University Library.** Manuscripts and Archives, Sterling Memorial Library, 120 High St., 1603A Yale Station, New Haven, CT 06520. Voice: 203 432 1775. Fax: 203 432 7231. The papers of the Morse Family includes about 190 letters by Morse, many letters to him, and other documents. Some 27 Morse letters are located in other Yale collections. Other relevant collections include papers of Lee de Forest, John Hays Hammond [wireless naval communications]; Mahlon Loomis [early radio, Loomis Aerial Telegraph Company]; Silliman Family (MS 74–1201); and Yale University Sheffield Scientific School Theses, 1895–1909 [academic papers on telegraphy and electrical communications].

2-B-6. Other Collections

0225   **Alexander Graham Bell Association for the Deaf Library.** 3417 Volta Place, Washington, DC 20007. Voice: 202 337 5220. Bell family materials (1870–1925), including business correspondence, laboratory notebooks, unpublished manuscripts. Access by appointment suggested.

0226   **American Antiquarian Society Library.** 185 Salisbury St., Worcester, MA 01609. Voice: 508 755 5221, Fax 508 754 9069. Collection of telegraph forms from pre-1870 companies.

0227   **American Institute of Physics.** Center for History and Philosophy of Physics, 335 East 45 St., New York, NY 10017. Voice: 212 685 1940. Collections include papers of J. Barton Hoag (MS 68–509) [television, ultra-shortwave radio, radio research]; Karl Lark-Horovitz (MS 68–511); Thomas Corwin Mendenhall (MS 65–1719) [Signal Corps, electrical engineering education]; George Washington Pierce (MS 68–515) [crystal oscillator, magnetostriction oscillator, wireless telegraphy, antennas, relays]; and George O. Southworth [Bell Telephone Laboratories, waveguide transmission].

0228   **Communications Workers of America.** Joseph A. Beirne Memorial Archives, 501 Third St. NW, Washington, DC 20001–2797. Voice: 202 728 2568, Fax 202 434 1201. Main collections include papers related to the union administrations of Joseph A. Beirne and Glenn Watts; the papers of Ray Hackney, the Telephone Guild of Wisconsin, and United Brotherhood of Telephone Workers related to the CWA's predecessor, the National Federation of Telephone Workers (NFTW); and the papers of the International Typographical Union (ITU).

0229   **Copley Press, Inc.** J.S. Copley Library, 1134 Kline St., P.O. Box 1530, La Jolla, CA 92038. Voice: 619 454 0411, ext. 341. Collections include selected very quotable papers (autograph letters, telegrams, and other manuscripts) related to telegraph and telephone by Morse, Bell, Abraham Lincoln, Mark Twain, Whitelaw Reid, Frank Fuller (Utah governor), and others.

0230   **The Electronics Museum of the Perham Foundation**, 101 First St., Suite 394, Los Altos, CA 94022, 408 734 4453 (temporary address and telephone). Most important collection is papers of Lee de Forest, 1873–1961 (MS 74–308), including de Forest's papers, scientific notebooks, scrapbooks, and other materials on wireless, radio, television, electronics; described in Earl G. Goddard's GUIDE TO FOOTHILL ELECTRONICS MUSEUM: MIRACLES IN TRUST (Perham Foundation, 1974). Other collections include Cledo Brunetti, 1936–1971 (MS 79–1685) [electronics and microelectronics in industry, aeronautics, military]; and Douglas Perham, 1887–1967 [early radio].

0231   **George C. Marshall Research Foundation Library.** P.O. Box 1600, Lexington, VA 24450–1600. Voice: 703 463 7103. Fax: 703 464 5229. Collections include papers of Marshall Andrews [Western Union Telegraph Company]; and Gilbert Sandford Vernam (MS 80–842) [communications patents, American Telephone and Telegraph Company].

0232   **Henry E. Huntington Library.** 1151 Oxford Rd., San Marino, CA 91108. Voice: 213 792 6141, Fax 818 405 0225. Collections include papers of James M. McClintock (Civil War Signal Corps]; Hiram Barney [telephone companies]; and William Heath Davis (MS 62–2145) [telegraph].

0233   **Institute of Electrical and Electronics Engineers, Inc.** Center for the History of Electrical Engineering, Piscataway, NJ 08855. Voice: 908 562 6835. The IEEE collection, 1880–date, includes corresponce, reports, minutes, and other records relating to history of IEEE and its predecessors, the American Institute of Electrical Engineers and the Institute of Radio Engineers.

0234   **Rockefeller Foundation.** Archive Center, 1133 Ave. of Americas, New York, NY 10036. Voice: 212 869 8500. Fax: 212 764 3468. Papers of the Foundation, 1912–1975, include records of support for electronics research and correspondence of Vannevar Bush, among others.

0235   **Sheldon Museum Research Center.** One Park St., Middlebury, VT 05753. Voice: 802 388 2117. Collections include papers of Henry Luther Sheldon, 1833–1907 [telegraph]; and Charles Linsley, 1784–1856 [Northern Telegraph Company].

0236   **Society of Wireless Pioneers, Inc.** Breniman Nautical-Wireless Library and Museum of Communications, P.O. Box 530, Santa Rosa, CA 95402. Voice: 707 542 0898. Maintains a special collection on development of wireless telegraphy (1850–date).

0237   **Southern California Library for Social Studies and Research.** 6120 Vermont Ave., Los Angeles, CA 90044. Voice: 213 759 6063. Library's 20,000 item pamphlet collections includes publications of the United Electrical, Radio, and Machine Workers (UE).

2-C. Selected Secondary Resources

2-C-1. Survey Histories

0238   Antebi, Elizabeth. **THE ELECTRONIC EPOCH**. New York: Van Nostrand Reinhold, 1983, 256 pp. Illustrated semi-technical history of 20th Century computer and communications electronics. Photos, diagrams, references, bibliography, index.

0239   Braun, Ernest, and Stuart Macdonald. **REVOLUTION IN MINIATURE: THE HISTORY AND IMPACT OF SEMICONDUCTOR ELECTRONICS**. Cambridge and New York: Cambridge University Press, 1982 (2nd ed.). 246 pp. From the work toward a transistor to integrated chips and beyond. Notes, tables, index.

0240   Brock, Gerald W. **THE TELECOMMUNICATIONS INDUSTRY: THE DYNAMICS OF MARKET STRUCTURE**. Cambridge, MA: Harvard University Press, 1983, 336 pp. A standard history of the industry to the eve of the AT&T divestiture, with insightful comments on both industry developments and policy trends. Notes, index.

0241   Dummer, G.W.A. **ELECTRONIC INVENTIONS AND DISCOVERIES**. Oxford, England: Pergamon, 1983 (3rd ed.), 233 pp. Chronological record of key inventions and patents across a variety of fields. Charts, tables, diagrams, index.

0242   Harlow, Alvin F. **OLD WIRES AND NEW WAVES: THE HISTORY OF THE TELEGRAPH, TELEPHONE, AND WIRELESS**. New York: Appleton-Century, 1936 (reprint: Arno Press, 1971), 548 pp. A still-useful standard history of telecommunications to the 1930s, equally enlightening on technology, companies, and key figures. Emphasis on the telegraph. Photos, index.

0243   Lubar, Steven. **INFOCULTURE: THE SMITHSONIAN BOOK OF INFORMATION AGE INVENTIONS**. Boston: Houghton Mifflin, 1993, 408 pp. Lushly illustrated history stressing technology, but also its applications and the companies behind the inventions. Photos, notes, index.

0244   Marvin, Carolyn. **WHEN OLD TECHNOLOGIES WERE NEW: THINKING ABOUT COMMUNICATIONS IN THE LATE NINETEENTH CENTURY**. New York: Oxford, 1988, 269 pp. Path-breaking social history based on changing technology—a kind of restrospective technology assessment. Photos, notes, index.

0245   Oslin, George P. **THE STORY OF TELECOMMUNICATIONS**. Macon, GA: Mercer University Press, 1992, 507 pp. Informal history especially good on Western Union where the author worked for many years. Photos, index.

0246   Winston, Brian. **MISUNDERSTANDING MEDIA**. New York: Oxford, 1986, 418 pp.. Title is misleading—this is a thoughtful analysis of the development of television, telephones, computers, and satellite communications, showing how they have *not* created a revolution due to society's control mechanisms. Notes, charts, index.

2-C-2. *Telegraph*

0247   Coe, Lewis. **THE TELEGRAPH: A HISTORY OF MORSE'S INVENTION AND ITS PREDECESSORS IN THE UNITED STATES**. Jefferson, NC: McFarland, 1993, 184 pp. A modern survey based largely on earlier seconary sources, this is useful for its longer view on a now-defunct technology. Photos, biographical sketches, bibliography, index.

0248   Marland, E.A. **EARLY ELECTRICAL COMMUNICATION**. London: Abelard-Schuman, 1964, 220 pp. Detailed study of early inventions and applications of telegraph technology—taking the story to the early telephone. Photos, bibliography, index.

0249   Reid, James D. **THE TELEGRAPH IN AMERICA: ITS FOUNDERS, PROMOTERS AND NOTED MEN**. New York: Derby Brothers, 1879 (reprinted by Arno Press in 1974), 846 pp. Invaluable for its detailed annotated directory-like treatment of early companies and people. Engravings.

0250   **REPORT OF THE TELEPHONE AND TELEGRAPH COMMITTEES OF THE FEDERAL COMMUNICATIONS COMMISSION IN THE DOMESTIC TELEGRAPH INVESTIGATION, DOCKET No. 14650, April 29, 1966.** Washington: GPO, 1966, 335 pp. Detailed the decline and virtual end of Western Union service and the reasons for it. Tables, charts (some fold-out), notes.

0251   Shiers, George, ed. **THE ELECTRIC TELEGRAPH: AN HISTORICAL ANTHOLOGY**. New York: Arno Press, 1977, ca 600 pp. Valuable collection of historical papers from 1859 to the mid-20th century, tracing technical changes in telegraphy. Photos, diagrams, maps, notes.

0252   Thompson, Robert Luther. **WIRING A CONTINENT: THE HISTORY OF THE TELEGRAPH INDUSTRY IN THE UNITED STATES 1832–1866**. Princeton, NJ: Princeton University Press, 1947, 544 pp. Intended as the first of a two-volume history, this is the only volume that appeared. It details the technical and corporate story to the formation of the Western Union near-monopoly. Notes, tables, sources, index.

2-C-3. *Telephone*

0253   **A HISTORY OF ENGINEERING AND SCIENCE IN THE BELL SYSTEM.** Whippany, NJ: Bell Telephone Laboratories, 1975–85, 7 vols. This details the premier private research lab, covering the past century—a model of self-published corporate technical history. Each volume focuses on specific area of work with well-researched original papers, heavily illustrated and well indexed. The individual volumes are not numbered—they are listed here in order of publication appearance:

   1 **THE EARLY YEARS (1925–1975)** edited by M.D. Fagan, 1975, 1,078 pp. Covers all aspects of technical work in AT&T to the formation of Bell Labs in 1926.

   2 **NATIONAL SERVICE IN WAR AND PEACE (1925–1975)** edited by M.D. Fagan, 1978, 757 pp. Includes detailed chapters on radar, electrical computers for fire control, communication, World War II, air defense, tactial and strategic defense systems, command and control, and operation of the Sandia National Labs.

3 **SWITCHING TECHNOLOGY (1925–1975)** edited by G.E. Schindler Jr., 1982, 639 pp. Evolution of electromechanical and manual switching, crossbar switches, preparing for full automation, direct distance dialing, electronic switching, private branch exchanges, etc.

4 **PHYSICAL SCIENCES (1925–1980)** edited by S. Millman, 1983, 674 pp. Chapters detail system research in physics (including the transistor, laser, etc.), and materials (semiconductors, fiber optics, etc.).

5 **COMMUNICATION SCIENCES (1925–1980)** edited by S. Millman, 1984, 521 pp. Includes chapters on the mathematical foundation of communication, acoustics, picture communication research, vacuum tube electronics, radio systems, fiber optic communications, switching, computer science, digital communications and behavioral science.

6 **TRANSMISSION TECHNOLOGY (1925–1975)** edited by E.F. O'Neill, 1985, 812 pp. Details such analog technologies as overseas and broadcast radio, coaxial cable, and UHF frequency work, microwave, submarine cable systems, satellites, mobile radio, and the advent of digital systems including the T1 carrier, fiber optic systems, and the technical and business environment for such technological work.

7 **ELECTRONICS TECHNOLOGY (1925–1975)** edited by F.M. Smits, 1985, 370 pp. The transistor, integrated chips, electron tubes, magnetic memories, capacitors and resistors and the like.

0254   **BRINGING INFORMATION TO PEOPLE: CELEBRATING THE WIRELESS DECADE**. Washington: Cellular Telecommunications Industry Association, 1993, 72 pp. Useful brochure issued by a trade association to review the first decade of U.S. cellular service.

0255   Chapuis, Robert J. **100 YEARS OF TELEPHONE SWITCHING: PART 1—MANUAL AND ELECTROMECHANICAL SWITCHING (1878–1960s).** Amsterdam: North-Holland, 1982, 482 pp. See next entry.

0256   _____, and Amos E. Joel, Jr. **ELECTRONICS, COMPUTERS AND TELEPHONE SWITCHING: 1960–1985**. Amsterdam: North-Holland, 1990, 595 pp. These volumes form an impressive international technical history of telephone switching, showing the relatively recent rise to importance of computers in that process. Photos, diagrams, charts, tables, notes, index.

0257   **DEPRECIATION: HISTORY AND CONCEPTS IN THE BELL SYSTEM**. New York: American Telephone & Telegraph, 1957, 154 pp. The sometimes complicated process is historically traced. Photos, charts, tables, bibliography. One of the few discussions of this topic available.

0258   Fischer, Claude S. **AMERICA CALLING: A SOCIAL HISTORY OF THE TELEPHONE TO 1940**. Berkeley: University of California Press,1992, 424 pp. Important analysis, based on a detailed study of three Bay-area California towns. Photos, notes, index.

0259   Pool, Ithiel de Sola. **FORECASTING THE TELEPHONE: A RETROSPECTIVE TECHNOLOGY ASSESSMENT OF THE TELEPHONE**. Norwood, NJ: Ablex, 1983, 162 pp. Well-indexed set of pre-1940 predictions on impact and social role. Index.

0260   Rhodes, Frederic Leland. **BEGINNINGS OF TELEPHONY**. New York: Harper, 1929 (reprinted: Arno Press, 1974), 261 pp. Detailed technical history of the instrument and related service through to about 1900, concentrating on AT&T developments. Photos, patent listing, source list, index.

0261   U.S. Department of Commerce, Bureau of the Census. **TELEPHONES AND TELEGRAPHS**. Washington: GPO, 1906-1939, title and pagination varies. Issued every five years to cover information from 1902-1937, these statistical reports are invaluable historical indicators of technical and economic development of the industry. The first couple of reports (for 1902 and 1907) include many illustrations as well. The series was originally part of the SPECIAL REPORTS sub-set of Census reports, but with issue for 1917 became the CENSUS OF ELECTRICAL INDUSTRIES series. Tables, charts.

0262   Walsh, J. Leigh. **CONNECTICUT PIONEERS IN TELEPHONY**. New Haven: Telephone Pioneers of America, 1950, 444 pp. One of the best of the breed, this is a TPA historical publication detailing telephone development in the state from its inception in the 1870s through the terrible floods of the late 1930s. Photos, notes, appendices, index.

0263   Wasserman, Neil H. **FROM INVENTION TO INNOVATION: LONG DISTANCE TELEPHONE TRANSMISSION AT THE TURN OF THE CENTURY**. Baltimore: Johns Hopkins University Press, 1985, 160 pp. The period before use of electronic amplification. Photos, diagrams, notes, index.

2-C-4. Wireless/Radio

0264   Aitken, Hugh G.J. **SYNTONY AND SPARK: THE ORIGINS OF RADIO**. New York: Wiley/Interscience, 1976 (reprint: Princeton University Press), 347 pp. See next entry.

0265   _____. **THE CONTINUOUS WAVE: TECHNOLOGY AND AMERICAN RADIO, 1900–1932**. Princeton, NJ: Princeton University Press, 1985, 588 pp. Detailed and well-written narrative on the key inventors and developmental trends of early radio, focusing on means of transmission and reception. Photos, notes, diagrams, index.

0266   Blake, G.G. **HISTORY OF RADIO TELEGRAPHY AND TELEPHONY**. London: Chapman & Hall, 1928 (reprinted by Arno Press, 1974), 425 pp. Highly detailed narrative and listing of early inventors and their patents. Photos, diagrams, 1,100-item reference list, index.

0267   Hancock, H.E. **WIRELESS AT SEA: THE FIRST FIFTY YEARS**. Chelmsford, England: Marconi International Marine, 1950 (reprinted by Arno Press, 1974), 260 pp. Detailed history of ship-to-shore wireless and radio services through World War II, emphasizing the role of the Marconi firm. Photos, index.

0268   Howeth, Captain L. S. **HISTORY OF COMMUNICATIONS-ELECTRONICS IN THE UNITED STATES NAVY**. Washington: GPO, 1963, 657 pp. Extremely detailed and important history of technical and policy developments of early wireless and radio (and related technologies). Photos, notes, index.

0269   Maclaurin W. Rupert. **INVENTION AND INNOVATION IN THE RADIO INDUSTRY**. New York: Macmillan, 1949 (reprinted by Arno Press, 1974), 303 pp. Definitive study concerning the role of patent control in development of a technology and then an industry. Photos, diagrams, notes, patent lists, bibliography, index.

0270   McNicol, Donald. **RADIO'S CONQUEST OF SPACE: THE EXPERIMENTAL RISE IN RADIO COMMUNICATION**. New York: Murray Hills Books, 1946 (reprinted by Arno Press, 1974), 374 pp. A radio engineer provides one of the best narratives on the people and inventions over more than a half century. Photos, diagrams, index.

0271   Shiers, George, ed. **THE DEVELOPMENT OF WIRELESS TO 1920**. New York: Arno Press, 1977, ca 600 pp. Valuable anthology of technical and historical articles from 1890 to 1920. Photos, diagrams, tables, notes.

0272   **YEAR-BOOK OF WIRELESS TELEGRAPHY AND TELEPHONY**. London: Marconi Company/The Wireless Press, 1913–1925 (annual, 13 volumes published), ca 1,000 pp. each. Historically important for its coverage of the point-to-point era of wireless, including extensive details on technical developments, laws, and existing shipboard and point-to-point stations around the world. Photos, maps, tables.

2-C-5. Satellite

0273   Brown, Martin P. Jr. **COMPENDIUM OF COMMUNICATION AND BROADCAST SATELLITES 1958 TO 1980**. New York: IEEE Press/John Wiley, 1981, 375 pp. Diagrams and photos detail the initial generation of communication satellites from all countries.

0274   Dunlap, Orrin E. Jr. **COMMUNICATIONS IN SPACE: FROM MARCONI TO MAN ON THE MOON**. New York: Harper & Row, 1970 (3rd ed.), 338 pp. Popular history of telecommunication stressing early communications satellites. Photos, index.

0275   Hudson, Heather. **COMMUNICATION SATELLITES: THEIR DEVELOPMENT AND IMPACT**. New York: Free Press, 1990, 338 pp. Very good combination of history, technological description and review of impact of both domestic and international satellites. Notes, glossary, bibliography, index.

0276   Magnant, Robert S. **DOMESTIC SATELLITE: AN FCC GIANT STEP TOWARD COMPETITIVE TELECOMMUNICATIONS POLICY**. Boulder: Westview, 1977, 296 pp. How changing technology led to a major policy breakthrough with the FCC's 1972 "open skies" open entry policy. Notes, diagrams.

0277   Smith, Delbert D. **COMMUNICATION VIA SATELLITE: A VISION IN RETROSPECT**. Leyden: Sijthoff, 1976, 335 pp. Relates the development of technology and policy for domestic and international satellites, both military and civilian. Notes, bibliography, index.

2-C-6. *Biography*

2-C-6-i. *Inventors*

0278   Aitken, William. **WHO INVENTED THE TELEPHONE?** London: Blackie, 1939, 196 pp. A stinging attack on Bell which argues others deserve the credit.

0279   Bruce, Robert V. **BELL: ALEXANDER GRAHAM BELL AND THE CONQUEST OF SOLITUDE**. Boston: Little, Brown, 1973, 564 pp. Definitive biography of the inventor. Photos, illustrations, notes, index.

0280   Carter, Samuel III. **CYRUS FIELD: MAN OF TWO WORLDS**. New York: Putnam, 1968, 380 pp. Details the life and impact of the key figure behind the pioneering mid-19th century trans-Atlantic submarine cables. Photos, notes, index.

0281   Harder, Warren J. **DANIEL DRAWBAUGH: THE EDISON OF THE CUMBERLAND VALLEY**. Philadelphia: Universiy of Pennsylvania Press, 1960, 228 pp. One of the many "might-have-been" inventors of the telephone is detailed. Photos, map, notes, bibliography, index.

0282   Jeffrey, Thomas E. **A GUIDE TO THE THOMAS A. EDISON PAPERS: A SELECTIVE MICROFILM EDITION**. Frederick, MD: University Publications of America, 1985, 2 parts. Detailed record series notes on Edison's patents, research notebooks, legal and finanacial records, correspondence, and scrapbooks, with index.

0283   Josephson, Matthew. **EDISON**. New York: McGraw-Hill, 1959, 511pp. Still the definitive treatment among a shelf-full on this quintessential American figure. Photos, notes, index.

0284   Maybee, Carleton. **AMERICAN LEONARDO: THE LIFE OF SAMUEL F.B. MORSE**. New York: Knopf, 1943, 435 pp. Still the definitive treatment of the telegraph system inventor's life, this won the Pulitzer Prize for biography the next year. Photos, references, index.

2-C-6-ii. *Other Important Figures*

0285   Paine, Albert Bigelow. **THEODORE N. VAIL: A BIOGRAPHY**. New York: Harper, 1921, 359 pp. The authorized life of the key management figure in the development of AT&T. Photos, index.

0286   Rhodes, Frederick Leland. **JOHN J. CARTY: AN APPRECIATION**. New York: Privately Printed, 1932, 280 pp. The long-time chief engineer of AT&T is detailed. Photos, patent list, index.

2-C-6-iii. *Collective Biography*

0287   Appleyard, Rollo. **PIONEERS OF ELECTRICAL COMMUNICATION**. London: Macmillan, 1930, 347 pp. Ten chapters on such figures as Maxwell, Wheatstone, Hertz, Heaviside, and others. Photos, index.

0288 Cortada, James W. **HISTORICAL DICTIONARY OF DATA PROCESSING BIOGRA-PHIES**. New York: Greenwood, 1987, 321 pp. Some 150 brief biographies include historically significant employees of AT&T, Bell Telephone Laboratories, RCA, IBM, and other electronics companies. Also describes oral history collections of Computer Museum, the Smithsonian's National Museum of American History (0127), MIT Archives (0177), University of Minnesota's Charles Babbage Institute, and others.

0289 Dunlap, Orrin E., Jr. **RADIO'S 100 MEN OF SCIENCE: BIOGRAPHICAL NARRA-TIVES OF PATHFINDERS IN RADIO, ELECTRONICS AND TELEVISION**. New York: Harper, 1944, 294 pp. Arranged by birth-date, this includes short (one to three page) lives and assessments of the work of key inventors.

0290 Hawks, Ellison. **PIONEERS OF WIRELESS**. London: Methuen, 1927 (reprinted by Arno Press, 1971), 304 pp. History of telecommunications to the 1920s as seen through the lives of key inventors and innovators. Photos, index.

0291 Potamian, Brother, and James J. Walsh. **MAKERS OF ELECTRICITY**. New York: Fordham University Press, 1909, 404 pp. Background of pioneers in the field in the 17th to 19th centuries. Drawings, notes.

0292 **TELECOMMUNICATION PIONEERS**. Long Island City, NY: Radio Engineering Laboratories, 1963, 61 pp. Twenty-seven pioneers are detailed, many from the 20th century. Index.

2-C-7. Company and Business History

2-C-7-i. General

0293 Borchardt, Kurt. **STRUCTURE AND PERFORMANCE OF THE U.S. COMMUNICA-TIONS INDUSTRY: GOVERNMENT REGULATION AND COMPANY PLANNING**. Cambridge, MA: Harvard Graduate School of Business Administration, 1970, 180 pp. With 0295, this is a valuable picture of the pre-competitive industry dominated by the unified Bell System of AT&T. Notes.

0294 **INTERNATIONAL DIRECTORY OF COMPANY HISTORIES**. Detroit: St. James Press, 1988–date, in progress. Exceptionally useful for informative histories of electronics, telecommunications, and information companies; global in scope. Through vol. 9 (1994), offers detailed sketches for more than 1,800 companies with annual sales over $500 million U.S. Basic data on headquarters, ownership, subsidiaries and affiliates, employees, sales; signed historical essays based on publically available sources, with sources for further reading. Entries in vols. 1–6 arranged in chapters for different industries: vol. 2, "Electronics" (INTEL, NEC, RCA, Siemens), "Entertainment—Broadcasting, Motion Pictures"; vol. 3, "Office Equipment and Personal Computers," "Information Technology" (Apple, Compaq, IBM, Xerox); vol. 4, "Publishing"; vol. 5, "Technology and Main Frame Computers," "Utilities—Telecommunications" (AT&T, Ameritech, Cable and Wireless, PTT Nederlands, NTT, Telefonica). Entries in vol. 7–date arranged alphabetically. Vols. 7–date include classified index to complete set to date: relevant listings under "Electrical and Electronics," "Telecommunications," "Utilities."

0295 Irwin, Manley R. **THE TELECOMMUNICATIONS INDUSTRY: INTEGRATION VS. COMETITION**. New York: Praeger, 1971, 223 pp. One of the earliest surveys of the industry, valuable today for its picture of the business just as competition was beginning. Notes, bibliography, index.

0296 Pleasance, Charles A. **THE SPIRIT OF INDEPENDENT TELEPHONY**. Johnson City, TN: Independent Telephony Books, 1989, 304 pp. The only current overall history of the non-Bell local exchanges. Notes, bibliography, index.

2-C-7-ii. Specific Firms

0297 Boettinger, H.M. **THE TELEPHONE BOOK: BELL, WATSON, VAIL AND AMERICAN LIFE**: 1876-1983. New York: Stern, 1983 (2nd ed.), 230 pp. Lavish pictorial history—much in color—of the telephone and AT&T until divestiture. Photos.

0298   Brooks, John. **TELEPHONE: THE FIRST HUNDRED YEARS**. New York: Harper & Row, 1976, 369 pp. A popular narrative history, written with some financial support from AT&T, this is the most approachable study available. Photos, notes, index.

0299   Cantelon, Philip C. **THE HISTORY OF MCI: THE EARLY YEARS, 1968-1988**. Dallas: Heritage Press, 1993, 719 pp. Privately-published narrative history which actually begins in 1963 and offers useful appendix on microwave history. Photos, appendices, index.

0300   Federal Communications Commission. **INVESTIGATION OF THE TELEPHONE INDUSTRY IN THE UNITED STATES**. 78th Cong, 1st Sess., House Document No. 340, 1939 (reprinted by Arno Press, 1974), 661 pp. The final report of a three-year FCC study of all aspects of AT&T history and operation offers the most detailed study of the company to that time. Tables.

0301   Garnet, Robert W. **THE TELEPHONE ENTERPRISE: THE EVOLUTION OF THE BELL SYSTEM'S HORIZONTAL STRUCTURE, 1876-1909**. Baltimore: Johns Hopkins Press, 1985. The central company and its regional affiliates—and how they grew. One of the AT&T-Hopkins series on telephone history. Photos, notes, index.

0302   Kahaner, Larry. **ON THE LINE: THE MEN OF MCI—WHO TOOK ON AT&T, RISKED EVERYTHING, AND WON!**. New York: Warner, 1986, 344 pp. A popular history of MCI, this breaks the history of the firm's first 25 years into eight different periods or "businesses," stressing the impact of legal actions and policies.

0303   Kleinfield, Sonny. **THE BIGGEST COMPANY ON EARTH: A PROFILE OF AT&T**. New York: Holt, Rinehart & Winston, 1981, 319 pp. Drawn from the authors series in the *New York Times*, this is a detailed sense of what before its breakup was the largest single business firm in the world. Index.

0304   Lipartito, Kenneth. **THE BELL SYSTEM AND REGIONAL BUSINESS: THE TELEPHONE IN THE SOUTH, 1877-1920**. Baltimore: Johns Hopkins University Press, 1989, 283 pp. One of the AT&T-Hopkins history of the telephone series, this details how AT&T expanded into ownership of local telephone firms, using the present territory of BellSouth as a case study example. Photos, tables, notes, index.

0305   McCarthy, Thomas E. **THE HISTORY OF GTE: THE EVOLUTION OF ONE OF AMERICA'S GREAT CORPORATIONS**. Stamford, CT: GTE, 1990, 224 pp. An in-house history, the only one available of the company which has long operated local telephone services and for a time controlled the Sprint long distance firm. Photos, notes, index.

0306   Reich, Leonard S. **THE MAKING OF AMERICAN INDUSTRIAL RESEARCH: SCIENCE AND BUSINESS AND GE AND BELL, 1876–1926**. New York: Cambridge University Press, 1985, 309 pp. Comparative study of early industrial research and development at two important electrical firms. Photos, notes, index.

0307   Smith, George D. **THE ANATOMY OF A BUSINESS STRATEGY: BELL, WESTERN ELECTRIC, AND THE ORIGINS OF THE AMERICAN TELEPHONE INDUSTRY**. Baltimore: Johns Hopkins University Press, 1985, 237 pp. Deals with the period up to 1915. Another of the AT&T-Hopkins series on the history of the telephone. Notes, photos, index.

2-C-7-iii. Unions and Trade Associations

0308   Brooks, Thomas R. **COMMUNICATIONS WORKERS OF AMERICA: THE STORY OF A UNION**. New York: Mason Charter, 1977, 257 pp. Details the rise of the CWA and its varied relationship with employers, chiefly AT&T. Index.

0309   Schacht, John N. **THE MAKING OF TELEPHONE UNIONISM, 1920–1947**. New Brunswick, NJ.: Rutgers University Press, 1985, 256 pp. Traces history of telephone workers' organizing efforts from company unions of 1920s, outlawed by Wagner Act, to founding of Communications Workers of America in 1947. Views movement for a centralized, bureaucratized union as response to AT&T's corporate labor policy.

0310   Secrest, James D. **EIA 50: ELECTRONIC INDUSTRIES ASSOCIATION: THE FIRST FIFTY YEARS**. Washington, DC: EIA, 1974, 248 pp. History of the main association of manufacturers. Photos.

2-C-8. Policy and Regulation

2-C-8-i. General

0311   **FIFTY YEARS OF *TELECOMMUNICATIONS REPORTS*.** Washington: Telecommunications Reports, 1985, 220 pp. Pulls together a weekly series published on the golden anniversary of the "yellow peril" weekly industry newsletter (1041), focusing primarily on regulation.

0312   Herring, James M., and Gerald C. Gross. **TELECOMMUNICATIONS ECONOMICS AND REGULATION**. New York: McGraw-Hill, 1936 (reprint: Arno Press, 1974), 544 pp. A classic early treatise on the 1934 Communications Act and what led to it, plus an assessment of the early FCC and its operations. Notes, appendices, index.

0313   Horwitz, Robert B. **THE IRONY OF REGULATORY REFORM: THE DEREGULATION OF AMERICAN TELECOMMUNICATIONS**. New York: Oxford, 1988, 448 pp. Interrelates the histories of electronic media and telecommunications to suggest comparatively little has really been deregulated. Notes, index.

0314   Kittross, John M. ed. **DOCUMENTS IN AMERICAN TELECOMMUNICATIONS POLICY**. New York: Arno Press, 1977 (2 vols.). Details early federal regulation of wireless and radio and includes (in second volume) the full Truman Board (1951) report and important documents on spectrum management.

0315   _____. **ADMINISTRATION OF AMERICAN TELECOMMUNICATIONS POLICY**. New York: Arno Press, 1980 (2 vols). Documents on development of FCC and radio facility licensing and (in second volume) studies of independent regulatory commissions (excerpts dealing with the FCC).

0316   Paglin, Max D., ed. **A LEGISLATIVE HISTORY OF THE COMMUNICATIONS ACT OF 1934**. New York: Oxford, 1989, 981 pp. All the key documents gathered in one place, plus current commentary essays—useful reference.

0317   President's Communications Policy Board. **TELECOMMUNICATIONS: A PROGRAM FOR PROGRESS**. Washington: GPO, 1951, 238 pp. Results of Korean War-era analysis of how telecommunications technology had changed and what government policies were needed to encourage both national security and private sector needs. Charts, notes.

0318   President's Task Force on Communications Policy. **FINAL REPORT**. Washington: GPO, 1969, 528 pp. The "Rostow" report assesses both media and telecommunication needs on the domestic and international levels as to what government structures and policies are needed.

0319   Stone, Alan. **PUBLIC SERVICE LIBERALISM: TELECOMMUNICATIONS AND TRANSITIONS IN PUBLIC POLICY**. Princeton, NJ: Princeton University Press, 1991, 296 pp. Incisive historical research reveals importance of government regulation in shaping the field even in its earliest years. Notes, index.

0320   Will, Thomas E. **TELECOMMUNICATIONS STRUCTURE AND MANAGEMENT IN THE EXECUTIVE BRANCH OF GOVERNMENT, 1900–1970**. Boulder, CO: Westview Press, 1978, 214 pp. Based on a dissertation, this details the approaches to federal regulation of and policy toward telecommunications outside of the FCC to 1970. Notes, appendices, bibliography.

2-C-8-ii. AT&T Divestiture and Aftermath (1974–94)

0321   Cole, Barry G., ed. **AFTER THE BREAK-UP: ASSESSING THE NEW POST-AT&T DIVESTITURE ERA**. New York: Columbia University Press, 1991, 480 pp. Collection of talks and papers based on a conference which examined all aspects of the industry in the first years after the breakup.

0322   Coll, Steve. **THE DEAL OF THE CENTURY: THE BREAKUP OF AT&T.** New York: Atheneum, 1986, 400 pp. Readable narrative on how and why AT&T was broken up, with excellent sense of key people and development of the issues in 1980–82. Index.

0323   Evans, David S., ed. **BREAKING UP BELL: ESSAYS ON INDUSTRIAL ORGANI-ZATION AND REGULATION**. New York: North-Holland, 1983, 298 pp. Economic analyses of the development of telephony, key aspects of the break-up case, and the impact of divestiture. Notes, charts, bibliography, index.

0324   Faulhaber, Gerald R. **TELECOMMUNICATIONS IN TURMOIL: TECHNOLOGY AND PUBLIC POLICY.** Cambridge, MA: Ballinger, 1987, 186 pp. Places the divestiture in historical context—why the change was almost inevitable given what had gone before. Notes, index.

0325   Henck, Fred W. and Bernard Strassburg. **A SLIPPERY SLOPE: THE LONG ROAD TO THE BREAKUP OF AT&T.** Westport, CT: Greenwood, 1988, 288 pp. Insightful history from 1934 by a pioneer journalist and a long-time chief of the FCC common carrier bureau. Index.

0326   Sterling, Christopher H., et al. eds. **DECISION TO DIVEST: MAJOR DOCUMENTS IN** *U.S. v. AT&T,* 1974–1988. Washington: Communications Press, 1986 (three volumes), with a fourth, DECISION TO DIVEST: THE FIRST REVIEW added in 1988. Introduction, user notes, and nearly 40 key decisions and filings (some complete, some excerpted) from Dept. of Justice, AT&T, and others.

0327   Stone, Alan. **WRONG NUMBER: THE BREAKUP OF AT&T** New York: Basic Books, 1989, 381 pp. Sympathetic to the industry viewpoint—argues the breakup was but the ordained end of a long process, related here. Notes, index.

0328   Temin, Peter with Louis Galambos. **THE FALL OF THE BELL SYSTEM**. New York: Cambridge University Press, 1987, 378 pp. Detailed, yet readable assessment based heavily on AT&T records in the antitrust case—perhaps the most complete to date. Notes, index.

0329   Tunstall, W. Brooke. **DISCONNECTING PARTIES—MANAGING THE BELL SYSTEM BREAK-UP: AN INSIDE VIEW**. New York: McGraw-Hill, 1985, 226 pp. How Bell dealt with the break-up as seen from the inside. Illustrations, index.

2-C-9. International

0330   Clark, Keith. **INTERNATIONAL COMMUNICATIONS: THE AMERICAN ATTITUDE**. New York: Columbia University Press, 1931 (reprinted: AMS Press, 1968), 261 pp. One of the earliest scholarly assessments, this offers sections on the Universal Postal Union, the International Telegraph Union, submarine cables, and the International Radio Union. Notes, index.

0331   Codding, George A., Jr. **THE INTERNATIONAL TELECOMMUNICATION UNION: AN EXPERIMENT IN INTERNATIONAL COOPERATION**. Leiden: E.J. Brill, 1952 (reprinted by Arno Press, 1972), 505 pp. Based on the author's dissertation, this is the definitive treatment of the ITU through World War II. Notes, definitions, bibliography, index.

0332   _____, and Anthony M. Rutkowski. **THE INTERNATIONAL TELECOMMUNI-CATION UNION IN A CHANGING WORLD**. Norwood, MA: Artech, 1982, 414 pp. The best overall survey of the ITU prior to its reorganization a decade later. Discusses history and the decision-making structure of the body. Notes, bibliography, index.

0333   Headrick, Daniel R. **THE INVISIBLE WEAPON: TELECOMMUNICATIONS AND INTERNATIONAL POLITICS 1851–1945**. New York: Oxford, 1991, 289 pp. Useful historical study of how nations worked (and sometimes competed) in the regulation and application of telecommunication, including international broadcasting. Notes, index.

0334   [Michaelis, Anthony]. **FROM SEMAPHORE TO SATELLITE**. Geneva: International Telecommunication Union, 1965, 343 pp. Illustrated official history of the ITU on its centenary. Photos, diagrams, bibliography.

0335   Schreiner, George Abel. **CABLE AND WIRELESS AND THEIR ROLE IN THE FOREIGN RELATIONS OF THE UNITED STATES**. Boston: Stratford Co., 1924, 269 pp. Much of this is devoted to undersea cables and the arrangments by which they are landed in various countries. Tables, appendices, index, map.

0336   Tomlinson, John D. **THE INTERNATIONAL CONTROL OF RADIOCOMMUNICATIONS**. Ann Arbor, MI: J.W. Edwards, 1945 (reprinted by Arno Press, 1979), 314 pp. History and development of worldwide agreements from 1903 until World War II. Notes, glossary, bibliography.

2-C-10   International Submarine Telegraphy

0337   Bright, Charles. **SUBMARINE TELEGRAPHS: THEIR HISTORY, CONSTRUCTION, AND WORKING**. London: Crosby, Lockwood and Son, 1898 (reprinted by Arno Press, 1974), 744 pp. Perhaps the single most important book detailing the first half century of what is often called "the grand Victorian technology". Lavishly illustrated with photos, charts, and maps (some fold-out), this discussion by a key engineer in major cable laying projects is an invaluable record.

0338   Brown, F.J. **THE CABLE AND WIRELESS COMMUNICATIONS OF THE WORLD**. London, Pitman, 1927, 148 pp. Useful update to title immediately above, relating early competitive stance between cable and wireless industries. Photos, maps, index.

0339   **THE CABLE AND WIRELESS COMMUNICATIONS OF THE WORLD 1924-1939**. London: Cable & Wireless Ltd., 1939, 282 pp. Collection of papers and talks which illustrate interwar developments. Photos, fold-out color map.

0340   Clarke, Arthur C. **HOW THE WORLD WAS ONE: BEYOND THE GLOBAL VILLAGE**. New York: Bantam, 1992, 296 pp. Updated and expanded version of the famed British science writer's VOICE ACROSS THE SEA, first published in 1958 and revised in a 2nd ed. In 1974. Details the full story of worldwide cable laying. Photos, maps, index.

0341   Coates, Vary T., and Bernard Finn. **A RETROSPECTIVE TECHNOLOGY ASSESSMENT: SUBMARINE TELEGRAPHY, THE TRANSATLANTIC CABLE OF 1866**. San Francisco: San Francisco Press, 1979, 264 pp. Fascinating study of social impact of faster international communication based on the success of the first lasting Atlantic telegraph cable. Charts, tables, notes.

0342   Dibner, Bern. **THE ATLANTIC CABLE**. Norwalk, CT: Burndy Library, 1959, 95 pp. Illustrated account of laying of first transoceanic cable, 1857–1866, with important bibliography of works on the Atlantic cable.

0343   Finn, Bernard S., ed. **DEVELOPMENT OF SUBMARINE CABLE COMMUNICATIONS**. New York: Arno Press, 1980, 2 vols, ca. 750 pp. An invaluable record of primary source documents and articles from 1858 to the 1970s. Photos, drawings, tables, maps.

0344   Garnham, Capt. S.A. and Robert Hadfield. **THE SUBMARINE CABLE: THE STORY OF THE SUBMARINE TELEGRAPH CABLE FROM ITS INVENTION DOWN TO MODERN TIMES—HOW IT WORKS, HOW CABLE-SHIPS WORK, AND HOW IT CARRIES ON IN PEACE AND WAR**. London: Sampson, Low, Marston Ltd., 1937, 242 pp. A 26-chapter survey of the technology with some comment on its use. Photos, appendix, index.

0345   Garratt, G.R.M. **ONE HUNDRED YEARS OF SUBMARINE CABLES**. London: Science Museum, 1950, 59 pp. Concise history emphasizing the cables, ships that laid them, and some detail on social and economic impact.

0346   Haigh, K.R. **CABLESHIPS AND SUBMARINE CABLES**. London: Adlard Coles, 1968, 418 pp. The most important source for historical information on all cable-laying vessels to the time of publication. Photos, maps, index.

2-C-11 Military Communications

0347   *Army Times*, editors of. **A HISTORY OF THE UNITED STATES SIGNAL CORPS**. New York: Putnam, 1961, 192 pp. Popular history in 12 chapters from the Civil War to pioneering satellite operations. Photos, bibliography, index.

0348   Bergen, John D. **MILITARY COMMUNICATIONS: A TEST FOR TECHNOLOGY— THE U.S. ARMY IN VIETNAM**. Washington: Center for Military History, 1986, 515 pp. Highly detailed survey in 20 chapters of changing communication technology from the 1950s into the mid-1970s, with one chapter on North Vietnamese developments. Maps, photos, notes, index.

0349   Lavine, A. Lincoln. **CIRCUITS OF VICTORY.** Garden City, NY: Doubleday Doran, 1921, 634 pp. Anecdotal history of American telephone facilities on the Western Front, 1917-19. Includes some discussion of both telegraph and wireless applications as well. Photos, maps.

0350   Marshall, Max L., ed. **THE STORY OF THE U.S. ARMY SIGNAL CORPS**. New York: Franklin Watts, 1965, 305 pp. Divided into two parts--half historical articles and half dealing with the Signal Corps as of the time of publication. Photos, appendices, index.

0351   Scheips, Paul J., ed. **MILITARY SIGNAL COMMUNICATIONS**. New York: Arno Press, 1980, two vols., ca 750 pp. Covers the history of Army use of the telegraph, telephone, and other means of electrical communications. Photos, diagrams, maps.

0352   Snyder, Thomas S., gen. ed. **THE AIR FORCE COMMUNICATIONS COMMAND: PROVIDING THE REINS OF COMMAND 1938-1981**. Scott Air Force Base, IL, I: AFCC Office of History, Scott Air Force Base, 1981, 231 pp. Well-illustrated history of the organization and technology of aircraft communications in the military. Photos, indexes.

0353   **THE SIGNAL CORPS—U.S. ARMY IN WORLD WAR II, THE TECHNICAL SERVICES**. Washington: GPO, as follows:
       1 Terrett, Dulany. **THE EMERGENCY.** 1956, 383 pp.
       2 Thompson, George Raynor, et al. **THE TEST** (December 1941 to July 1943), 1957, 621 pp.
       3 Thompson, George Raynor and Dixie R. Harris. **THE OUTCOME** (Mid-1943 through 1945), 1966, 720 pp. The series is a well-documented and detailed assessment of both technology development and applications, as well as Signal Corps operations during the War. Photos, tables, maps, index.

0354   Woods, David. **A HISTORY OF TACTICAL COMMUNICATION TECHNIQUES**. Orlando, FL: Martin-Marietta Corp, 1965 (reprinted by Arno Press, 1977), 300 pp. Only history of its type—cutting across the years from messengers and signal flags to electrical means. Photos, diagrams, bibliography.

2-C-12. Collecting Telecommunications Devices

0355   Dooner, Kate E. **TELEPHONES ANTIQUE TO MODERN**. West Chester, PA: Schiffer, 1992. 175 pp. See next entry.

0356   _____. **TELEPHONE COLLECTING: SEVEN DECADES OF DESIGN**. Atglen, PA: Schiffer, 1993, 128 pp. These two color-illustrated albums provide a good sense of the growing market in old telephones and related objects, here with many photos and guides for collectors.

0357   Povey, P.J., and R.A.J. Earl. **VINTAGE TELEPHONES OF THE WORLD**. Piscataway, NJ: IEEE Service Center, 1988, 202 pp. A British volume useful for its details on European telephone instruments in what is otherwise a narrative history of the instrument.

3

Technology

Chapter 3 includes bibliographies and other basic research resources on telecommunication technology. Powell's outstanding SELECTIVE GUIDE TO LITERATURE ON TELECOMMUNICATIONS (0369) and Hurt's INFORMATION SOURCES IN SCIENCE AND TECHNOLOGY (0364) describe telecommunications' highly technical literature as well as more accessible resources. Likewise, ENGINEERING INDEX (0378) analyzes the field's highly technical research journals as well as other journals that are fundamental for policy studies. Sections 3-B and 3-C include primary and secondary resources and national and international agencies and organizations for research on U.S. telecommunications patents and national and international telecommunications standards. Emphasis here is identifying the few basic tools for research in very complex and highly technical areas. We include selected secondary handbooks and studies of transmission modes (wire/cable, broadband, satellite, fiber optic, and mobile) and spectrum management that provide accessible explanations of these areas.

3-A. Bibliographic Resources

3-A-1. Bibliographies

0358   Ardis, Susan. **A GUIDE TO THE LITERATURE OF ELECTRICAL AND ELECTRONICS ENGINEERING.** Littleton, CO: Libraries Unlimited, 1987, 190 pp. Standard guide to advanced engineering and other technical reference as well as other information resources. Covers patents, standards, computer software, data compilations, product catalogs, company data books, document delivery sources, etc. Telecommunications materials described throughout, particularly handbooks (pp. 80–82) and reference texts (pp. 95–97). Author/title and subject indexes.

0359   **BIBLIOGRAPHIC GUIDE TO COMPUTER SCIENCE.** Boston: G.K. Hall, 1988–date, annual. (Bibliographic Guides). Subject bibliography identifying items added to collections of MIT and Stanford University libraries. Entries under "Telecommunications."

0360   **BIBLIOGRAPHIC GUIDE TO TECHNOLOGY.** New York: G.K. Hall, 1987–date, annual. "Comprehensive annual subject bibliography" cumulating holdings of New York Public Library and Library of Congress. Useful for international coverage.

See 0910. **BIBLIOGRAPHY OF PUBLICATIONS IN COMMUNICATION SATELLITE TECHNOLOGY AND POLICY.**

0361   Coolidge, Andrea, and Teresa Gorman. **ANNOTATED BIBLIOGRAPHY ON TELEMATICS: THE TECHNOLOGIES OF TELECOMMUNICATIONS, COMPUTERS, AND INFORMATION.** Monticello, IL: Vance Bibliographies, 1982, 29 pp. (Public Administration Series, P–883). About 90 classified, annotated references to mostly books, reports, and government documents; sections for technologies, impacts, applications, international, research, policy issues. Covers such topics as electronic funds transfer, environmental health, and labor and employment.

0362   DeMerritt, Lynne. **SITING OF POWER LINES AND COMMUNICATION TOWERS: A BIBLIOGRAPHY ON THE POTENTIAL EFFECTS OF ELECTRIC AND MAGNETIC FIELDS.** Chicago: Council of Planning Librarians, 1990, 27 pp. (CPL Bibliography 257). Pages 8–12 include 30 annotated references to mostly EPA and local environmental protection documents on communications facilities and RF radiation in relation to ANSI guidelines. Appendix lists EMF monitoring organizations.

0363   Haynes, David, ed. **INFORMATION SOURCES IN INFORMATION TECHNOLOGY.** Munich: G.K. Saur, 1990, 350 pp. Evaluates basic bibliographies and reference works, organizations, trade and scholarly journals, patent and standards sources, U.S., U.K., and other national government documents related to particular information technologies, their applications, and information technology policy and regulation.

0364   Hurt, C.D. **INFORMATION SOURCES IN SCIENCE AND TECHNOLOGY.** Englewood, CO: Libraries Unlimited, 1994 (2nd ed.), 412 pp. Classified annotated guide to reference resources (abstracts and indexes, dictionaries, handbooks, directories, biographical sources, serials, etc.). Listings for science and technology in general and for specific fields. Sources for telecommunications identified throughout, especially in chapters for "Electrical engineering" and "Multidisciplinary resources." Subject index offers little help under "telecommunications"; instead, users must investigate under more specialized terms, including "Electro-optical communications," "Fiber optics," "Radio engineering," etc. Author/title and subject indexes.

0365   **LIBRARY HI TECH BIBLIOGRAPHY.** Ann Arbor, MI: Pierian Press, 1986–date, annual. Volumes include 10–12 extensive annotated bibliographies of recent literature on variety of high technology topics; for example, vol. 1: "Cable Television," "Telecommuting"; vol. 2: "Interactive Video"; vol. 5: "Electronic Bulletin Boards"; vol. 7: "Cable Television," HDTV, "Productivity and Information Technology," "Telecommunications Regulation and Deregulation," "Telecommuting"; vol.8: "Open Systems Interconnection."

0366   Maier, Ernest L. **SATELLITE TO EARTH TV RECEPTION: WITH PARTICULAR EMPHASIS ON PERSONAL HOME RECEPTION.** Monticello, IL: Vance Bibliographies, 1981, 9 pp. (Public Administration Series, P–815). About 100 references to popular and trade journal articles, 1973–1980, on privately-owned earth stations.

0367   Moore, C.K., and K.J. Spencer. **ELECTRONICS: A BIBLIOGRAPHICAL GUIDE.** London: Macdonald, 1961, 411 pp. See next entry.

0368   _____. **ELECTRONICS: A BIBLIOGRAPHICAL GUIDE—2.** New York: Plenum, 1965, 369 pp. Comprehensive guide to technical research literature, covering reference works, journals, monographs, and articles. 1961 vol., frequently cited as ELECTRONICS I, includes 2,877 annotated entries from 1945–1959 in classified arrangement of 68 chapters. 1965 vol. includes 2,880 entries from 1959–1964 in 70 chapters. Relevant chapters include "Applications of Electronics to Science, Medicine, and Industry," "Radio-Communication," "Telemetering," and others. Author and subject indexes.

0369   Powell, Jill H. **SELECTIVE GUIDE TO LITERATURE ON TELECOMMUNICA-TIONS.** Washington: American Society for Engineering Education, 1993, 15 pp. (Engineering Literature Guides, number 15). Very informative descriptions of technological literature of telecommunication, covering selected bibliographies, printed and electronic indexes and abstracts, dictionaries and encyclopedias, handbooks, introductory texts and key works, journals, conference proceedings, directories, and standards sources.

0370   Sanchez, James Joseph. **TELETEXT SYSTEMS: A SELECTIVE, ANNOTATED BIBLI-OGRAPHY.** Monticello, IL: Vance Bibliographies, 1987, 9 pp. (Public Administration Series, P–2265). About 25 items from trade journals, ERIC documents, with international focus.

0371   _____. **VIDEOTEX SYSTEMS: A SELECTIVE, ANNOTATED BIBLIOG-RAPHY.** Monticello, IL: Vance Bibliographies, 1987, 17 pp. (Public Administration Series, P–2262). About 60 annotated items; international focus (Prestel, Minitel, etc.).

3-A-2. Selected Abstracting, Indexing, and Electronic Database Services

0372   ACM GUIDE TO COMPUTING LITERATURE. New York: Association of Computing Machinery, 1977–date, annual. Online version MATHSCI available from Dialog, ESA/IRS. CD-ROM version MATHSCI available from American Mathematical Society. Formerly COMPUTING REVIEWS: BIBLIOGRAPHY AND SUBJECT INDEX OF CURRENT COMPUTING LITERATURE (1968–1976). Indexes major computing journals, conference papers, reports, and theses. Access by authors, keywords, products, sourcees.

0373   APPLIED SCIENCE AND TECHNOLOGY INDEX. New York: H. W. Wilson, 1958–date, monthly. Online version available from Wilsonline, OCLC: covers October 1983–date, updated twice weekly. CD-ROM version available from Silver Platter, H. W. Wilson: covers 1983–date, updated quarterly. Author and subject indexing of English-language technical and scientific journals in computer technology and applications and electrical and telecommunications engineering. Covers most prominent telecommunication and new communication technologies journals. Useful subject headings include "Computer networks," "Optical communication systems," "Signal processing," "Telecommunication," "Telephone."

0374   COMPUTER ABSTRACTS. Bradford, England: MCB University Press, 1957–date, monthly. Annotated entries in classified arrangement: relevant chapters include "Communications and Networks," "Applications—Communications"; "Human-Computer Interaction"; and others. Useful subject index: entries under "Telecommunications," "Teleoperation," "Telerobots," etc.

0375   COMPUTER AND CONTROL ABSTRACTS. London: Institution of Electrical Engineers/ Piscataway, NJ: Institute of Electrical and Electronics Engineers, 1966–date, monthly. Formerly CONTROL ABSTRACTS (1966–1968). Series C of SCIENCE ABSTRACTS (1898–date). Online version INSPEC available from BRS, Dialog, STN, Orbit. CD-ROM version INSPEC available from IEE/IEEE. Covers journals, reports, dissertations, conference papers issued worldwide in computer and control engineering. Author, subject, bibliography, book, and conference indexes. See next entry.

0376   ELECTRICAL AND ELECTRONICS ABSTRACTS. London: Institution of Electrical Engineers/Piscataway, NJ: Institute of Electrical and Electronics Engineers, 1966–date, monthly. Formerly ELECTRICAL ENGINEERING ABSTRACTS (1941–1966). Series B of SCIENCE ABSTRACTS (1898–date). Online version INSPEC available from BRS, Dialog, STN, Orbit, OCLC. CD-ROM version INSPEC available from IEEE. Covers books, journals, reports, dissertations, and conference papers worldwide on all aspects of electrical engineering. Author, subject, bibliography, book, and conference indexes. Electronic version INSPEC (cumulative SCIENCE ABSTRACTS) offers important historical coverage. KEY ABSTRACTS: TELECOMMUNICATIONS (1975–date, monthly) is one of several subsets of the INSPEC database issued by IEEE/IEE for current awareness.

0377   ELECTRONICS AND COMMUNICATIONS ABSTRACTS JOURNAL. Riverdale, MD: Cambridge Scientific Abstracts, 1967–date, monthly. Online version available from BRS, ESA/ IRS. Formerly ELECTRONICS ABSTRACTS JOURNAL. Covers trade information and research published in books, journals, reports, dissertations, conference papers, and patents related to electronic systems, physics, circuits and devices, and communications.

0378   ENGINEERING INDEX. New York: Engineering Information, Inc., 1934–date, monthly. Online version COMPENDEX PLUS available from BRS, Dialog, Knowledge Index, Orbit Search Service, Tech Data: coverage varies (1970–date on Dialog, Orbit), updated monthly. CD-ROM version COMPENDEX PLUS available from Engineering Information, Inc.: covers 1970–date, updated monthly. Most important index to telecommunication's technical research literature. Abstracts, with subject and author indexes, over 4,500 international scholarly, technical, and professional journals, technical reports, monographs, conference proceedings, and other publications. Covers telecommunication and new communication technologies' core journals, such as TELECOMMUNICATION JOURNAL (1152), TELECOMMUNICATIONS POLICY (1123), and TELEMATICS AND INFORMATICS (1107), as well as major non-U.S. engineering and scientific titles. Listings arranged by subject headings and subheadings based on SHE: SUBJECT HEADINGS FOR ENGINNERING (New York: Engineering Information, Inc., 1972–date, annual). Major headings include "Communication satellites," "Telecommunications," and "Telephones," with subheadings for "Developing Countries," "Economic and Sociological Effects," and proper names. Separate subject index cumulates headings, sub-headings, names, acronyms, popular terms, etc., and cross references main listings. Indexes for author and authors' affiliations. Abridged version, CONCISE ENGINEERING AND TECHNOLOGY INDEX, available online as ENGINDEX/FS on OCLC.

0379   **GOVERNMENT REPORTS ANNOUNCEMENTS AND INDEX**. Springfield, VA: National Technical Information Service, 1975–date, monthly with annual cumulations. Online version NTIS available from BRS, Dialog, Orbit, STN: covers 1964–date, updating varies. CD-ROM version NTIS available from Silver Platter: covers 1983–date, updated quarterly. Annually abstracts more than 70,000 reports, datafiles, proceedings, guides, manuals, and other items related to U.S. government-sponsored research. Full bibliographic citations arranged by subject category and subcategory; for example, "Communication—Policies, Regulations, & Studies." Individual and institutional/corporate author, keyword, report and accession number indexes.

0380   **INDEX TO IEEE PUBLICATIONS**. New York: Institute of Electrical and Electronics Engineers, 1973–date, monthly. CD-ROM version IEEE/IEE PUPLICATIONS ONDISC available from University Microfilms: covers 1988–date, updated monthly. Formerly CUMULATIVE INDEX TO I.R.E PUBLICATIONS (1951–1958), TWENTY YEAR CUMULATIVE INDEX (1951–1971) and INDEX TO IEEE PERIODICALS (1971–1973). Author and subject index to about 300 IEEE publications, including journals, technical transactions, conference papers, and standards. Access under familiar "tele"-words as well as others like "telecontrol," "telemetering," and "telewriting." "Telecommuting" cross referenced to "home working."

0381   **SCIENCE CITATION INDEX**. Philadelphia: Institute for Scientific Information, 1961–date, bimonthly with annual cumulations. Online version SCISEARCH available from Dialog, Data-Star, others: coverage varies, updated weekly. CD-ROM version available from ISI: current coverage only, updated quarterly. Author, cited reference, and keyword indexes for some 4,500 journals.

3-A-3. Directories

See 0496. **CORPORATE TECHNOLOGY DIRECTORY**.

0382   **DIRECTORY OF AMERICAN RESEARCH AND TECHNOLOGY**. New Providence, NJ: R.R. Bowker, 1965–date, annual. Formerly INDUSTRIAL RESEARCH LABORATORIES OF THE UNITED STATES (1920–1964). CD-ROM version available from Bowker Electronic Publishing in SCI-TECH REFERENCE PLUS: covers current edition, updated annually. Subtitle: "Organizations active in product development for business." 28th ed. for 1994 (1993). Profiles private and public companies listed alphabetically under parent company. Entries give brief information, identify numbers and academic disciplines of Ph.D.s, technicians. Geographic, personnel, and "R&D Classification" (subject) indexes. Relevant listings under "Communication Systems and Equipment," "Information Systems," Radio Communication," "Satellites," "Telecommunications," "Telephone Systems and Equipment."

3-A-4. Dictionaries

See Section 1-D.

3-B. Patents

3-B-1. U.S. Patent Office

0383   **U.S. Department of Commerce, Patent and Trademark Office**. 2021 Jefferson Davis Highway, Arlington, VA 22202. Patent or Trademark information, voice: 703 557 4636. The repository of the nation's patent history (though some older records are in the National Archives, 0120). Holds as public records all patents (primarily in paper files, though computerization project is making headway).

0384   U.S. Department of Commerce. Patent and Trademark Office. **CASSIS**. [CD-ROM.] Washington: GPO, 1969–date, bimonthly. Searching U.S. patent information typically requires using the U.S. Patent and Trademark Office's INDEX TO THE U.S. PATENT CLASSIFICATION, MANUAL OF CLASSIFICATION, and PATENT CLASSIFICATION DEFINITIONS (all listed seperately below); in combination, these cross reference classification and subclassification numbers and descriptive titles to provide an index to the OFFICIAL GAZETTE OF THE UNITED STATES PATENT AND TRADEMARK OFFICE, a weekly abstracting and announcing service that gives details for patents granted (see 0368, 0387, 0390, and 0388). CASSIS, or the "Classification and Search Support Information System," designed for use in numerically-organized patent collections (especially patent depository libraries), simplifies research by providing current classification information and electronic versions of Patent Office publications. CASSIS includes two complementary parts: CASSIS/CLSF indexes all U.S. patents from 1790 to the present by classification and subclassification and patent number (including some 10,000 X-numbered patents issued in 1790–1836 without numbers); and CASSIS/BIB, covering 1969 to the present, accesses patents by classifications, descriptive titles, names of patentees and assignees, dates, geographic locations, and most usefully, keywords in titles and abstracts.

0385   _____. **DIRECTORY OF PATENT DEPOSITORY LIBRARIES.** Washington: Patent and Trademark Office, 1986–date, annual. Identifies public and academic depository libraries.

0386   _____. **INDEX TO THE U.S. PATENT CLASSIFICATION**. Washington: GPO, 1988–date, revised annually. Alphabetical list of everyday-language terms used to describe inventions are cross referenced to classification and subclassification numbers.

0387   _____. **MANUAL OF CLASSIFICATION**. Washington: GPO, 1988–date, annual. Lists all classification and subclassification numbers and descriptive titles, with details on scope of classifications and notes referencing related classifications.

0388   _____. **OFFICIAL GAZETTE OF THE UNITED STATES PATENT AND TRADEMARK OFFICE: PATENTS.** Washington: GPO, 1975–date, weekly. The Patent Office's abstracting and announcing service, the bulk of each weekly issue numerically arranges in three parts reflecting the U.S. classification system (general and mechanical, chemical, and electrical), with a separate section for design patents. Indexes for classifications, patentees, and other data. A similar separate weekly publication covers trademarks: OFFICIAL GAZETTE OF THE UNITED STATES PATENT AND TRADEMARK OFFICE: TRADEMARKS (Washington: GPO, 1975–date). Previous titles include: OFFICIAL GAZETTE OF THE UNITED STATES PATENT OFFICE (1872–1971) and OFFICIAL GAZETTE OF THE UNITED STATES PATENT OFFICE: PATENTS (1971–1975). OFFICIAL GAZETTE is supplemented by the annual INDEX OF PATENTS ISSUED FROM THE UNITED STATES PATENT AND TRADEMARK OFFICE (Washington: GPO, 1974–date), arranged includes separate lists of patentees and subjects of inventions numerically arranged in classification system order. ANNUAL REPORT OF THE COMMISSIONER OF PATENTS (Washington: GPO, 1840–1965) provides descriptions of patents granted from 1849 to 1871. Information on U.S. patents granted before 1849 is identified in the variety of indexes to Congressional records, particularly INDEX TO THE SERIAL SET (Washington, DC: Congressional Quarterly, 1975–1980), as well as EARLY UNNUMBERED UNITED STATES PATENTS, 1790–1836: INDEX AND GUIDE TO THE MICROFILM EDITION (New Haven, CT: Research Publications, 1980).

0389   _____. **PATENT AND TRADEMARK OFFICE COLLECTION OF HISTORICAL AND INTERESTING U.S. PATENTS IN CELEBRATION OF OUR NATION'S BICENTENNIAL.** Washington: GPO, 1976, 1 microfilm reel. A very convenient packaging of full texts of patents of historical significance, including those of Morse, William Robinson, Edison, Bell,

Emile Berliner, Francis Blake, Almon B. Stowger, Valdemar Poulsen, Lee de Forest, Edwin H. Armstrong, Carl R. Englund, Philo Farnsworth, V. K. Zworykin, and Robert N. Noyce.

0390 _____. **PATENT CLASSIFICATION DEFINITIONS.** Washington: GPO, 1987–date, microfiche regularly revised. Describes subclassifications more completely and identifies alternative classifications.

0391 _____. **TELEPHONE DIRECTORY.** Washington: Patent and Trademark Office, 1989–date, annual. Identifies telephone numbers of sections responsible for telecommunication classifications; this useful directory information, as well as U.S. Patent Classifications, is reprinted in Richard C. Levy's INVENTING AND PATENTING SOURCEBOOK: HOW TO SELL AND PROTECT YOUR IDEAS (Detroit: Gale, 1990).

3-B-2. Secondary Resources on Patents

0392 Bertin, Gilles Y., and Sally Wyatt. **MULTINATIONALS AND INDUSTRIAL PROPERTY: THE CONTROL OF THE WORLD'S TECHNOLOGY.** Brighton: Harvester, 1988, 177 pp. Discusses tensions between inventors and imitators and developed and developing countries in high technology, including telecommunications and elctronics.

0393 Bowie, Norman E. **UNIVERSITY-BUSINESS PARTNERSHIPS: AN ASSESSMENT.** Lanham, MD: Rowman & Littlefield, 1994, 287 pp. Discusses business versus values issues of corporate-sponsored academic research and development in advanced technologies. Particular emphasis on technology transfer at MIT in telecommunications and information technologies, with historical documents on patents and licensing agreements.

0394 Kraeuter, David W. **RADIO AND TELEVISION PIONEERS: A PATENT BIBLIOG-RAPHY.** Metuchen, NJ: Scarecrow, 1992, 319 pp. Identifies by patent number, date, and descriptive title more than 3,000 patents for Armstrong, de Forest, Dolbear, Farnsworth, Du Mont, Loomis, Marconi, Sarnoff, and many others. Companion volume to Kraeuter's BRITISH RADIO AND TELEVISION PIONEERS: A PATENT BIBLIOGRAPHY (Metuchen, NJ: Scarecrow, 1993).

0395 Marzouk, Tobey B. **PROTECTING YOUR PROPRIETARY RIGHTS IN THE COMPUTER AND HIGH TECHNOLOGY INDUSTRIES.** Washington: Computer Society Press, 1988, 208 pp. Practical guide to protecting industrial property rights.

0396 McKnelly, Michele, and Johanna Johnson. **PATENTS AND TRADEMARKS: A BIBLI-OGRAPHY OF MATERIALS AVAILABLE FOR SELECTION.** Washington: Patent Depository Library Association, 1989, 63 pp. Guide to major information resources about U.S. patents.

0397 Office of Technology Assessment and Forecast. **TELECOMMUNICATIONS.** Washington: GPO, 1984, 261 pp. (Patent Profiles). Analyzes over 48,000 U.S. patents issued from 1963 to 1983 in telephony, light wave communications, multiplex communications, analog carrier wave communications, digital and pulse communications, television and facsimile transmission, and telemetry. Corresponds to TELECOMMUNICATIONS: PATENT PROFILES MICROFICHE SUPPLEMENT (August 1984) which gives U.S. patent numbers by technology area for analyzed patents. Other volumes in Patent Profiles series include MICROELECTRONICS (1981 AND 1983) as well as volumes on robotics, biotechnology, etc.

0398 Warshofsky, Fred. **THE PATENT WARS: THE BATTLE TO OWN THE WORLD'S TECHNOLOGY.** New York: Wiley, 1994, 298 pp. Views world courts as battlefields in war for intellectual property rights and patent clustering and litigation as financial survival tools. Suggests liberalizing and streamlining U.S. patent system as means to enhance and protect American innovations. Focuses on patent battles of Apple v. Microsoft, Borland v. Lotus, as well as AT&T, IBM, Motorola, and NEC. Index.

3-C. Technical Standards

3-C-1. Resources for Identifying Telecommunication Standards

0399   American National Standards Institute. **CATALOG OF AMERICAN NATIONAL STANDARDS.** New York: ANSI, 1923–date, annual with supplements. Formerly UNITED STATES OF AMERICA STANDARDS INSTITUTE CATALOG and variant titles. Annual sales catalog (with ordering information) consisting of "A comprehensive subject index to all currently approved American National Standards" and listing by ANSI and other standard designations and titles, such as ANSI/ AIAA, ANSI/ICEA, and ANSI/TIA/EIA. Useful headings include "Information systems—data communication" and "telecommunication." Available on CD-ROM in cooperation with IHS (see 0401).

0400   Folts, Harold C. **McGRAW-HILL'S COMPILATION OF OPEN SYSTEMS STANDARDS.** New York: McGraw-Hill, 1990 (4th ed.), 6 vols. ANSI, ISO, EIA, CCITT, ECMA, and FED and FIPS standards.

0401   **INDEX AND DIRECTORY OF INDUSTRY STANDARDS.** Englewood, CO: Information Handling Services, 1989–date, annual. Continues INDEX AND DIRECTORY OF U.S. INDUSTRY STANDARDS (1983–1989). Produced by Information Handling Services, 15 Inverness Way E., P. O. Box 1154, Englewood, CO 80150. Voice: 303 790 0600, 800 525 7052. Fax: 303 790 0686. Online version IHS INTERNATIONAL STANDARDS & SPECIFICATIONS available from Dialog, updated weekly. CD-ROM version WORLDWIDE STANDARDS SERVICE available from IHS, updated bimonthly. The most convenient comprehensive index to a broad range of current standards. 1992 ed. (in 5 vols.) identifies 123,000 standards documents from 409 organizations; claims to cover 90% of the world's most referenced standards as well as U.S. military specifications. Vols. 1 and 2 cover U.S. standards, vols. 3–5, international, non-U.S. national standards. Subject indexes, numeric indexes (by standards organizations), ANSI number concordances, and directories of organizations. Listings offer helpful scope notes, cross references, and subheadings. Data for standards includes issuing body, designation number, title, notice of ANSI approval, pagination, parent title (if issued as part of book or series), date of latest revision or reapproval, and Department of Defense adoption information. Acronyms dictionary and addresses of organizations. Electronic versions offer access by titles, subjects, descriptors, standard designations, U.S. Federal Supply Product codes (for example, 5805 for "Telephone and telegraph equipment"), and sources or issuing organizations.

0402   International Electrotechnical Commission. **CATALOGUE OF IEC PUBLICATIONS.** Geneva, Switzerland: IEC, 1983–date, annual. Lists publications of IEC, an international electrical and electronics standardization organization. Complements coverage of ISO CATALOGUE (0403).

0403   International Organization for Standardization. **ISO CATALOGUE.** Geneva, Switzerland: ISO, 1989–date, annual. Lists all ISO standards by subject and number, with alphabetical index and list of obsolete standards. Relevant listings under "Open Systems Interconnection" and "Information processing systems."

0404   **OMNICOM INDEX OF STANDARDS FOR DISTRIBUTED INFORMATION AND TELECOMMUNICATION SYSTEMS: 1989 (4th) EDITION.** Ed. Harold C. Folts. Vienna, VA: Omnicom; New York: McGraw Hill Information Services, 1989, 906 pp. 1st ed. published in 1986. Bibliographic citations with abstracts and ordering information for standards from 22 national and international organizations (ANSI, IEEE, ISO, CCITT, Department of Defense, EC, UK, Canada, Denmark, Germany, Japan, Hungary, Ireland, Netherlands, Saudi Arabia, Sweden, Australia, Finland) related to Open Systems Interconnection (OSI), ISDN, and data communications. Indexed by subjects/key words and organizations.

0405   **QRIS: QUICK REFERENCE TO IEEE STANDARDS.** New York: Institute of Electrical and Electronics Engineers, 1980–date, annual. Index of terms in IEEE and ANSI standards published by IEEE, with contents of each standards document.

0406   Ruffner, James A., and Linda R. Musser, comps. **UNION LIST OF TECHNICAL REPORTS, STANDARDS, AND PATENTS IN ENGINEERING LIBRARIES.** Washington: American Society for Engineering Education, 1992, 34 pp. (Engineering literature guides, number 13). Produced by the Engineering Libraries Division of the American Society for Engineering Education. Identifies standards collections in broad selection of academic and technical libraries, with details on statuses of holdings (depository, complete, partial, etc.) and access services (interlibrary loan, photocopying, etc.).

0407   Stallings,   William.   **HANDBOOK   OF   COMPUTER-COMMUNICATIONS STANDARDS.** Carmel, IN: Sams, 1990 (2nd ed.), 3 vols. Comprehensive reference for "technology, implementation, design, and application issues" of computer-communications standards (p. xi). Indended to serve as a companion for standards documents.

0408   _____. **NETWORKING STANDARDS: A GUIDE TO OSI, ISDN, LAN, AND MAN STANDARDS.** Reading, MA: Addison-Wesley, 1993, 646 pp. A practical tutorial and reference guide that tries to "translate standards documents into a form more palatable for readers." Focuses on implementation of computer-communications architechture and protocols.

3-C-2. Major U.S. Standards Organizations, Collections, and Delivery Services

0409   **American National Standards Institute** (ANSI). 11 West 42nd St., New York, NY 10036. Voice: 212 642 4900. Fax: 212 302 1286. ANSI membership includes more than 1,200 professional and technical societies, trade associations, and companies. ANSI approves standards developed and voluntarily submitted by trade, technical, and other organizations; coordinates and assigns ANSI numbers; represents U.S. interests in nongovernmental standards fora with ISO and IEC; and supplies ANSI and other national and international standards. Joint Telecommunications Standards Coordinating Committee coordinates telecommunication standards-making organizations. Supplies ANSI and other standards, including CCITT and ITU Recommendations; accepts orders by phone, fax, and mail. Publishes CATALOG OF AMERICAN NATIONAL STANDARDS (0399); ANSI REPORTER (1967–date), biweekly newsletter, reports on standards news and activities of its membership.

See 0555. **Alliance for Telecommunications Industry Solutions.**

0410   **Cleveland Public Library,** Business and Science Bldg., 2nd Floor, North, 325 Superior Ave., Cleveland, OH 44114-1271. Voice: 216 623 2932 (reference), 216 623 2901 (photoduplication). Perhaps the U.S.'s most comprehensive standards collections open to the public. Holds standards from more than 40 organizations and all U.S. government agencies (military, federal, or DODISS), including ANSI, EIA, IEEE, and IPC; particularly important historical files of obsolete ANSI, ASTM, NFPA, and other standards.

See 0559. **Electronic Industries Association** (EIA).

See 0782. **Federal Communications Commission.**

0411   **Global Engineering Documents.** 2805 McGraw Ave., P.O. Box 19539, Irvine, CA 92714. Voice: 714 261 1455, 800 854 7179. Fax: 714 261 7892. Important standards vendor. Supplies hard copies of all standards and specifications included in IHS INTERNATIONAL STANDARDS & SPECIFICATIONS (0401) on demand. Claims to maintain the "largest technical specifications library in the world."

See 0575. **Institute of Electrical and Electronics Engineers** (IEEE).

See 0773.2. **International Telecommunications Advisory Committee.**

See 0770.5. **National Center for Standards and Certification Information** (NCSCI).

See 0770.5. **National Institute of Standards and Technology** (NIST).

See 0770.6. **National Technical Information Service** (NTIS).

See 0761. **Office of Technology Assessment** (OTA).

0412   **Omnicom, Inc.** 115 Park St., SE, Vienna, VA 22180-4607. Phone: (703) 281-1135. Fax: (703) 281-1505, (800) 666-4266. Supplies OSI standards including CCITT, ECMA, and ISO JTC 1.

3-C-3. Secondary Resources on Standards

0413   **INFORMATION TECHNOLOGY STANDARDS: THE ECONOMIC DIMENSION.** Paris: Organization for Economic Cooperation and Development, 1991, 108 pp. Changing concepts for such standards, standards policy, traditional and new players, etc. Tables, notes.

0414   Macpherson, Andrew. **INTERNATIONAL TELECOMMUNICATION STANDARDS ORGANIZATIONS.** Norwood, MA: Artech House, 1990, 317 pp. An excellent handbook describing the composition and operation of international, national, and industry telecommunication standard-making bodies, with phone and fax numbers.

0415   Office of Technology Assessment. **GLOBAL STANDARDS: BUILDING BLOCKS FOR THE FUTURE.** Washington: GPO, 1992, 114 pp. Discusses the role and process of standard setting in general, comparing American and European experience. Diagrams, tables, notes, index.

0416   Ricci, Patricia. **STANDARDS: A RESOURCE AND GUIDE FOR IDENTIFICATION, SELECTION, AND ACQUISITION.** Woodbury, MN: Pat Ricci Enterprises, 1992 (2nd ed.), 336 pp. The best guide to resources for standards research, including detailed directories of the world's national and international standards organizations (noting affiliations with ANSI, ISO, etc.), U.S. government standards agencies, international standards libraries and information centers (identifying collections of current and obsolete standards), standards vendors, and consultants (indexed by subject); bibliography of standards resources; acronym and master indexes.

0417   **"STANDARDS: THEIR GLOBAL IMPACT,"** *IEEE Communications* 32:1 (January 1994). Special issue with nine papers reviewing the status of worldwide standards-setting bodies and policies.

0418   Wallenstein, Gerd. **SETTING GLOBAL TELECOMMUNICATION STANDARDS: THE STAKES, THE PLAYERS & THE PROCESS.** Norwood, MA: Artech House, 1990, 256 pp. Emphasizes the role of global entities in worldwide agreement on technical standards. Charts, glossary, bibliography, index.

3-D. Selected Secondary Resources

3-D-1. Handbooks/Surveys

0419   Bertsekas, Dimitri, and Robert Gallagher. **DATA NETWORKS.** Englewood Cliffs, NJ: Prentice Hall, 1992 (2nd ed.), 556 pp. Textbook for advanced students covering layered network architecture, point-to-point protocols, delay models, multi-access communications, and flow control.

0420   Conrad, James W. **HANDBOOK OF COMMUNICATIONS SYSTEMS MANAGEMENT.** Boston, MA: Auerbach, 1994 (3rd ed.), 845 pp. Covers communication systems planning, architecture and standards, local and metropolitan networks, systems management, issues and trends. Updated by yearly supplement.

0421   Dordick, Herbert S. **UNDERSTANDING MODERN TELECOMMUNICATIONS.** New York: McGraw-Hill, 1986, 324 pp. Non-technical survey of technologies, applications and impacts. Diagrams, glossary, index.

0422   Elliot, Douglas F., ed. **HANDBOOK OF DIGITAL SIGNAL PROCESSING: ENGINEERING APPLICATIONS.** San Diego, CA: Academic Press, 1987, 999 pp. Basic and advanced engineering approaches to transforms, FIR filters, IIR digital filters, spectral analysis, fast discrete transforms, time delay estimation, and mechanization of digital systems processors.

0423   Fortier, Paul J. **HANDBOOK OF LAN TECHNOLOGY.** New York: McGraw-Hill, 1992 (2nd ed.), 732 pp. Comprehensive overviewof new developments in LAN software services and innovations in LAN hardware. Emphasis on LAN design, digital communications, technology topologies and requirements, security, management, and applications.

0424   Freeman, Roger L., ed. **REFERENCE MANUAL FOR TELECOMMUNI-CATIONS ENGINEERING.** New York: Wiley, 1994 (2nd ed.), 2,308 pp. Massive guide in 31 subject-divided chapters on all aspects of the technology of telecommunications, updated from a 1985 edition. Designed for telecommunications systems engineers. Extensive bibliographies. Indexed.

0425   Goldman, James E. **APPLIED DATA COMMUNICATIONS: A BUSINESS-ORIENTED APPROACH.** New York: Wiley, 1994, 643 pp. Detailed and well-illustrated review of the equipment and services markets for data communication. Effective use of two-color printing throughout. Diagrams, charts, tabls, references by chapter, index.

0426   Grant, August E., ed. **COMMUNICATION TECHNOLOGY UPDATE.** Newton, MA: Focal Press, 1994 (3rd ed.), 389 pp. This annual is extremely usesful for its tightly organized chapters, each dealing with a specific technology or family of technologies, and each written to cover background, recent developments (last couple of years) and factors to watch in the future. Tables, charts, notes, glossary.

0427   Green, James Harry. **THE BUSINESS ONE IRWIN HANDBOOK OF TELECOMMU-NICATIONS.** Homewood, IL: Business One Irwin, 1992 (2nd ed.), 1,119 pp. Thirty-four chapters explore all aspects of service and equipment. Appendices, glossary, bibliography, index.

0428   Heldman, Robert K. **FUTURE TELECOMMUNICATIONS: INFORMATION APPLI-CATIONS, SERVICES & INFRASTRUCTURE.** New York: McGraw-Hill, 1993, 234 pp. See next entry.

0429   _____. **GLOBAL TELECOMMUNICATIONS: LAYERED NETWORKS, LAYERED SERVICES.** New York: McGraw-Hill, 1992, 390 pp. Useful surveys of current and coming technologies written in largely non-technical fashion. Diagrams, charts, glossary, index.

0430   Inglis, Andrew F., ed. **ELECTRONIC COMMUNICATIONS HANDBOOK.** New York: McGraw-Hill, 1988, 1 vol. (various pagings). Part I covers transmission and switching technologies; part II covers planning, design, and contruction of electronic communications systems.

0431   Martin, James. **TELECOMMUNICATIONS AND THE COMPUTER.** Englewood Cliffs, NJ: Prentice-Hall, 1990 (3rd ed.), 720 pp. A standard work with material useful to those with or without technical training. Charts, photos, diagrams, notes, index.

0432   Minoli, Daniel. **TELECOMMUNICATIONS TECHNOLOGY HANDBOOK.** Norwood, MA: Artech, 1991, 772 pp. Broad survey of 1990s' technologies and applications. Diagrams, index.

0433   Mirabito, Michael M.A. **THE NEW COMMUNICATIONS TECHNOLOGIES.** Newton, MA: Focal Press, 1994 (2nd ed.), 223 pp. Basic text of 11 chapters, each devoted to a specific technology or issue. Illustrations, notes, index.

0434   Noll, A Michael. **INTRODUCTION TO TELECOMMUNICATION ELECTRONICS.** Norwood, MA: Artech, 1988, 359 pp. Excellent introductory survey which assumes little prior knowledge. Diagrams, index.

0435   Office of Technology Assessment. (U.S. Congress) **ADVANCED NETWORK TECHNOL-OGIES.** Washington, DC: GPO, 1993, 79 pp. Describes six prototype gigabit networks funded federally that may provide keys to the future broadband standard network.

0436   Pecar, Joseph A., et al. **THE McGRAW-HILL TELECOMMUNICATIONS FACTBOOK.** New York: McGraw-Hill, 1993, 373 pp. As cover says, "a readable guide to planning and acquiring products and services." Eighteen chapters appear on telecommunication fundamentals, voice services, data services, integrated digital networks, and other technologies, standards, and systems. Diagrams, glossary, bibliography, index.

0437   Rosner, Roy D, **SATELLITES, PACKETS, AND DISTRIBUTED TELECOMMUNI-CATIONS: A COMPENDIUM OF SOURCE MATERIALS.** Belmont, CA: Lifetime Learning Publications, 1984, 628 pp. Technical and managerial handbook for distributed networks. Glossary.

0438   Stallings, William. **ADVANCES IN INTEGRATED SERVICES DIGITAL NETWORKS (ISDN) AND BROADBAND ISDN.** Los Alamitos, CA: IEEE Computer Society Press, 1992, 255 pp. Chapters on ISDN, ISDN protocols and network architecture, frame relay, broadband ISDN, asynchronous transfer mode, and synchronous optical network/synchronous digital hierarchy.

0439   _____. **DATA AND COMPUTER COMMUNICATIONS.** New York: Macmillan, 1994 (4th ed.), 875 pp. Highly technical graduate-level textbook covering principles and topics on technology and architechture of data and computer communications, with some emphasis on developments in broadband ISDN. Glossary and bibliography.

3-D-2. Transmission Modes

3-D-2-i. Wire/Cable (including Submarine Cables)

0440   Clark, Martin P. **NETWORKS AND TELECOMMUNICATIONS: DESIGN AND OPERATION.** New York: Wiley, 1991, 635 pp. Basics of networks in diagram and text, including setting them up and maintenance. Bibliographies, diagrams, glossary, index.

0441   National Telecommunications and Information Administration. **1990 WORLD'S SUBMARINE TELEPHONE CABLE SYSTEMS.** Washington: GPO, 1990, 515 pp. Details (including maps and specifications) of all then-operating cables around the globe. Photos.

0442   Noll, A. Michael. **INTRODUCTION TO TELEPHONES AND TELEPHONE SYSTEMS.** Norwood, NJ: Artech House, 1991 (2nd ed.), 244 pp. Ten chapter introductory survey. Photos, diagrams, glossary, bibliography, index.

0443   Rey, R.F., ed. **ENGINEERING AND OPERATIONS IN THE BELL SYSTEM.** Murray Hill, NJ: AT&T Bell labroatories, 1983 (2nd ed.), 884 pp. Detailed description of technical basics of telephony as practiced by the pre-divestiture AT&T. Photos, diagrams, references, glossary, index.

0444   Schenck, Herbert H. **CABLESHIP CHARACTERISTICS.** Washington, DC: Undersea Cable Engineers, Inc. 1978, 287 pp. Full details on all then-operating cableships of the world as well as full information on how cables are layed and repaired.

0445   Walters, Rob. **COMPUTER TELEPHONE INTEGRATION.** Norwood, MA: Artech, 1993, 352 pp. Details on integrating the two technologies, including applications and case studies. Glossary, index.

0446   Wrobel, Leo A. **DISASTER RECOVERY PLANNING FOR TELECOMMUNICA-TIONS.** Norwood, MA: Artech, 1990, 112 pp. Just that—all the things that can go wrong and how best to prevent them—or recover from them. Illustrations, glossary.

3-D-2-ii. Broadband Networks

0447   Davidson, Robert P. **BROADBAND NETWORKING ABCs FOR MANAGERS: ATM, BISDN, CELL/FRAME RELAY TO SONET.** New York: Wiley, 1994, 176 pp. Details on the variety of new service and technology options being developed for data networking. Charts, tables, glossary, index.

0448   Handel, Rainer, and Manfred N. Huber. **INTEGRATED BROADBAND NETORKS: AN INTRODUCTION TO ATM-BASED NETWORKS.** Workingham, England; Reading, MA: Addison-Wesley, 1991, 230 pp. Presents planning, development, implementation, and sales of telecommunications networks and terminals. Handel and Huber are telecommunications experts with Siemens.

0449   Heldman, Robert K. **TELECOMMUNICATIONS MANAGEMENT PLANNING: ISDN NETWORKS PRODUCTS AND SERVICES.** Blue Ridge Summit, PA: TAB Books, 1987, 734 pp. See next entry.

0450   _____. **ISDN IN THE INFORMATION MARKETPLACE.** Blue Ridge Summit, PA: TAB Books, 1988, 333 pp. These two volumes from the then technical director for US West detail the kinds of services available and how best to initiate and manage same. Diagrams, tables, bibliography, index.

0451   Muller, Nathan J. **INTELLIGENT HUBS.** Norwood, NJ: Artech, 1993, 336 pp. Hubs and hub interfaces and their application in various apsects of network management. Several vendor-specific chapters. Diagrams, glossary, bibliography, index.

0452   Ramteke, Timothy. **NETWORKS.** Englewood Cliffs, NJ: Prentice Hall Career and Technology, 1994, 482 pp. Broadscale survey of networks and networking: equipment and services, voice networking, data networking, and local area networks. Diagrams, notes, index.

0453   Rutkowski, Anthony M. **INTEGRATED SERVICES DIGITAL NETWORKS.** Norwood, MA: Artech, 1985, 324 pp. While now dated, this remains useful for its discussion of the decision-making forums and their work. the process of developing ISDN standards, and U.S. policymaking concerns. Diagrams, bibliography, glossary, index.

0454   Wright, David. **BROADBAND: BUSINESS SERVICES, TECHNOLOGIES AND STRATEGIC IMPACT.** Norwood, NJ: Artech, 1993, 476 pp. The technologies and related business services are detailed. Final chapters assess application of broadband to the insurance, publishing and health care industries. Charts, tables, bibliography, glossary, index.

3-D-2-iii. Satellites

0455   Binkowski, Edward S. **SATELLITE INFORMATION SYSTEMS.** Boston: G.K. Hall, 1988, 213 pp. Eight chapters designed for nontechnical readers, describing the industry and major issues facing users and carriers. Notes, bibliography, index.

0456   Gordon, Gary D., and Walter L. Morgan. **PRINCIPLES OF COMMUNICATIONS SATELLITES.** New York: Wiley, 1993, 553 pp. Advanced graduate-level textbook on communications satellites. Focuses on link power budgets and spacecraft technology.

0457   Jansky, Donald M., and Michael C. Jeruchim. **COMMUNICATIONS SATELLITES IN THE GEOSTATIONARY ORBIT.** Dedham, MA: Artech, 1987 (2nd ed.), 633 pp. First 100 pages focus on policy while remainder deals with technical and interference-reduction questions. Tables, charts, index.

0458   Long, Mark, ed. **WORLD SATELLITE ALMANAC: THE GLOBAL GUIDE TO SATELLITE TRANSMISSION & TECHNOLOGY.** Winter Beach, FL: MLE Inc., 1992 (regularly revised), 1,050 pp. A standard directory combining articles and directory information. Diagrams, tables.

0459   Rees, David W.E. **SATELLITE COMMUNICATIONS: THE FIRST QUARTER-CENTURY OF SERVICE.** New York: Wiley, 1990, 329 pp. This is not the history the title suggests, but rather offers descriptions of current satellite projects and systems here and abroad. Most material focuses on applications of satellite services. Photos, diagrams, appendices, bibliography, glossary, index.

0460   Wood, James. **SATELLITE COMMUNICATIONS & DBS SYSTEMS.** Stoneham, MA: Focal Press, 1992, 279 pp. Basic and non-technical survey of systems and applications here and abroad. Photos, diagrams, glossary, bibliography, index.

3-D-2-iv. Fiber Optic

0461   Chaffee, C. David. **THE REWIRING OF AMERICA: THE FIBER OPTICS REVOLUTION.** Orlando: Academic Press, 1988, 241 pp. Non-technical narrative tells the story of fiber optic development and expanding application in many different aspects of life. Photos, index.

See 0520. Federal Communications Commission, **FIBER DEPLOYMENT UPDATE**.

0462   Geller, Henry. **FIBER OPTICS: AN OPPORTUNITY FOR A NEW POLICY?** Washington: Annenberg Washington Program, 1992, 75 pp. Policy-driven discussion of ways in which expanding fiber installations may undo old "bottleneck" based policies.

0463   Reed, David M. **RESIDENTIAL FIBER OPTIC NETWORKS: AN ENGINEERING AND ECONOMIC ANALYSIS**. Norwood, MA: Artech, 1992, 354 pp. Focus is on so-called "last mile" connections and combines analysis of technological base with economic options. Charts, tables, bibliography, glossary, index.

0464   Yates, Robert K., et al. **FIBER OPTICS AND CATV BUSINESS STRATEGY**. Norwood, MA: Artech, 1990, 159 pp. Designed for cable system operators, this is a detailed guide on the options available and how best to accomplish each. Diagrams, index.

   3-D-2-v. Mobile Services

0465   **A COMPETITIVE ASSESSMENT OF THE U.S. CELLULAR RADIOTELEPHONE INDUSTRY.** Washington: GPO (International Trade Administration, U.S. Dept. of Commerce), 1988. Good overview of the mobile cellular business here and abroad, though now considerably dated. (See title just below.)

0466   Balston, D.M., and R.C.V. Marcario, eds. **CELLULAR RADIO SYSTEMS**. Norwood, MA: Artech, 1993, 373 pp. Details American, European and Japanese systems and their varied technologies and technical standards. Photos, diagrams, acronym list, index.

0467   Calhoun, George. **DIGITAL CELLULAR RADIO**. Norwood, MA: Artech, 1988, 448 pp. Good material on development of mobile cellular systems,focusing on problems with analog systems and how digital technology will prove beneficial. Diagrams, tables, notes, index.

0468   _____. **WIRELESS ACCESS AND THE LOCAL TELEPHONE NETWORK**. Norwood, MA: Artech, 1992, 595 pp. Includes discussion of personal communication systems and other options and how they may best be interconnected with the backbone national network. Diagrams, notes, index.

0469   Jagoda, A. and M. de Villepin. **MOBILE COMMUNICATIONS**. New York: Wiley, 1993, 180 pp. Mobile services and products are reviewed in chapters focusing on technology, policy, equipment, and major players in the field--comparing American and European systems.

0470   Mehrota, Asha. **CELLULAR RADIO: ANALOG AND DIGITAL SYSTEMS**. Norwood, MA: Artech, 1994, 460 pp. Detailed and well-illustrated chapters on the technology of both American and foreign systems, highlighting the transition to digital.

0471   Muller, Nathan J. **WIRELESS DATA NETWORKING**. Norwood, MA: Artech, 1995, 346 pp. Compares and contrasts cellular and other wireless systems, including regional operations of RBOC firms.

0472   Paetsch, Michael. **THE EVOLUTION OF MOBILE COMMUNICATIONS IN THE U.S. AND EUROPE: REGULATION, TECHNOLOGY, AND MARKETS**. Norwood, MA: Artech, 1993, 417 pp. History and current operations of the different systems amidst the transition from analog to digital operation. Diagrams, tables, index.

0473   Underwood, Geoffrey, et al. **INFORMATION TECHNOLOGY ON THE MOVE: TECHNICAL AND BEHAVIOURAL EVALUATIONS OF MOBILE TELECOMMUNICA-TIONS**. New York: Wiley, 1993, 245 pp. Based in large part on British experience, this book assesses road transport informatics—the use of telecommunication technology to speed traffic and make it safer. Diagrams, charts, glossary, index.

3-D-3. Spectrum Management

0474   Glatzer, Hal. **WHO OWNS THE RAINBOW? CONSERVING THE RADIO SPECTRUM**. Indianapolis, IN: Howard W. Sams, 1984, 302 pp. Non-technical discussion of why spectrum and its management are important. Diagrams, notes, index.

0475   Kobb, Bennett Z. **SPECTRUM GUIDE: RADIO FREQUENCY ALLOCATIONS IN THE UNITED STATES 30 MHz-300 GHz**. Falls Church, VA: New Signals Press, 1994, 311 pp. Useful reference with narrative description of spectrum allocations and actual uses of the VHF and EHF frequency bands.

0476   Levin, Harvey J. **THE INVISIBLE RESOURCE: USE AND REGULATION OF THE RADIO SPECTRUM**. Baltimore: Johns Hopkins Press, 1971, 432 pp. Though dated, this remains a standard work that combines discussion of the physics of spectrum propagation, spectrum management, options for different kinds of spectrum management, etc. Tables, charts, notes, index.

0477   McGillem, Clare D., and William P. McLauchlan, **HERMES BOUND: THE POLICY AND TECHNOLOGY OF TELECOMMUNICATIONS**. West Lafayette, IN: Purdue University Office of Publications, 1978, 284 pp. While dated, this remains one of the more insightful discusions of the politics often underlying spectrum decisions. References, glossary, index.

0478   National Telecommunication and Information Administration. **U.S. SPECTRUM MANAGEMENT POLICY: AGENDA FOR THE FUTURE**. Washington: GPO, 1991. 200 pp. Good discussion of the interface between technical limits and policymaking concerns. Notes, diagrams.

0479   _____. **PRELIMINARY SPECTRUM REALLOCATION REPORT**. Washington: GPO, 1994, ca. 150 pp. Details projected plans to reallocate 200 MHz of government spectrum to private uses.

0480   Office of Technology Assessment. **THE 1992 WORLD ADMINISTRATIVE RADIO CONFERENCE: TECHNOLOGY AND POLICY IMPLICATIONS**. Washington: GPO, 1993, 190 pp. Useful summary—first by this agency in more than a decade. Discusses technology and the international context for spectrum management. Photos, diagrams, notes, index.

4

Industry and Economics

This chapter details resources on economic and business aspects of telecommunications. In addition to including bibliographies, indexes, and other resources that identify telecommunications' business/trade and economics and management/scholarly literature, we give particular attention to statistical resources published by both government and industry. By far the most useful guidance to telecommunications' statisitcs is provided by STATISTICAL MASTERFILE (0513). Researchers with access to its accompanying microfiche collections should consider themselves very fortunate indeed. Also of considerable importance is COMPACT D/SEC DISCLOSURE (0545), containing full texts of annual reports and other federally-required company documents. Section 4-E identifies a small selection of major trade and industry associations, professional organizations, and public interest groups (we also list directories of telecommunications consultants) because, we believe, their offices can provide supplemental industry statistics. Selected secondary resources include industry surveys, economic analyses, and management studies.

4-A. Bibliographic Resources

4-A-1. Bibliographies

0481   Du Charme, Rita. **BIBLIOGRAPHY OF MEDIA MANAGEMENT AND ECONOMICS**. Minneapolis: Media Management and Economics Research Center, University of Minnesota, 1988 (2nd ed.), 131 pp. Unannotated, classified listings. Relevant items under "Cable," "Technology and Telecommunications" (pp. 63–66), and "Video and Videotext" (36 items), and elsewhere.

0482   Fildes, Robert, David Dews, and Syd Howell. **A BIBLIOGRAPHY OF BUSINESS AND ECONOMIC FORECASTING**. Westmead, England: Gower, 1981, 424 pp. Computer-produced listing of about 4,000 journal articles from 1971–1978. Relevant listings under "Utilities—Telephone," "Services—Computers," "Production and Mining—Electronics," "Technology," and others.

0483   Fuld, Leonard M. **THE NEW COMPETITOR INTELLIGENCE: THE COMPLETE RESOURCE FOR FINDING, ANALYZING, AND USING INFORMATION ABOUT YOUR COMPETITORS**. New York: John Wiley, 1995, 482 pp. Guide to printed, electronic, and organizational resources and strategies for their uses in research on companies and industries. International in scope, especially strong for Japanese high technologies. Directory, subject index.

0484   **HOW TO FIND INFORMATION ABOUT COMPANIES**. "1993/1994 edition." Washington: Washington Researchers, 1993, 3 vols. Detailed descriptions of company information sources: volume 1 covers published printed and electronic resources; federal, state, and local documents, organizations, trade and professional associations, etc.; volume 2 offers advice and case studies on how to research companies and corporate culture, plants and facilities, research and development, mangement and labor, products and services, finances, strategic plans; volume 3 index.

0485   Middleton, Karen P., and Meheroo Jussawalla. **THE ECONOMICS OF COMMUNICATION: A SELECTED BIBLIOGRAPHY WITH ABSTRACTS**. New York: Pergamon, 1981, 249 pp. Classified annotated listing of 386 items (books, popular and scholarly articles, conference papers, NTIS documents, etc.) on general and specific aspects of communication's economics, including competition and monopolies, demand, pricing, and impact of communications on economic systems. Author and subject indexes: relevant listings under "Telecommunications," and other headings.

0486   Miller, E. Willard, and Ruby M. Miller. **UNITED STATES TRADE—INDUSTRIES: A BIBLIOGRAPHY.** Monticello, IL: Vance Bibliographies, 1991, 25 pp. (Public Administration Series, P–3059). Pages 11–17 list about 70 unannotated references on U.S. international trade in semiconductors, home electronics, fiber optics, communications equipment, magnetic recorders, electronics equipment, and computers.

0487   Popovich, Charles J., and M. Rita Costello. **DIRECTORY OF BUSINESS AND FINANCIAL INFORMATION SERVICES.** Washington: Special Libraries Association, 1994 (9th ed.), 471 pp. Detailed descriptions of printed and electronic business, economic, and industry information sources. Title, publisher, and subject indexes. Very convenient for identifying numerous telecommunications newsletters available from Dialog, Dow Jones, LEXIS/NEXIS (0728), NewsNet, and other electronic services. Relevant publications listed under "Communication," "Electronics," "Telecommunications," "Telephone," and others.

0488   Roess, Anne C. **PUBLIC UTILITIES: AN ANOTATED GUIDE TO INFORMATION SOURCES.** Metuchen, NJ: Scarecrow Press, 1991, 340 pp. Of interest here primarily for section IV on the telephone industry (35 pages, about 180 items) as well as the general material on public utilities. Index.

0489   Towse, Raymond J. **HIGH TECHNOLOGY INDUSTRY, INDUSTRIALIZATION AND LOCATION: A BIBLIOGRAPHY AND GEOGRAPHICAL REVIEW.** Monticello, IL: Vance Bibliographies, 1988, 99 pp. (Public Administration Series, P–2496). Over 1,000 unannotated references on use of technology and innovation (R&D, production, and management) in selected high technology industries, including semiconductor, electronics, communications equipment, and computer industries.

4-A-2. Selected Abstracting, Indexing, and Electronic Database Services

0490   **ABI/INFORM.** [Machine-readable data.] Ann Arbor: UMI/Data Courier, 1971–date, weekly and monthly. Online service available from American Library Association, BRS, Dialcom, Dialog, Executive Telecom Systems, Knowledge Index, Mead Data Central, Orbit Search Services, Tech Data. Online coverage varies (1970–date on BRS), updating varies. CD-ROM version ABI/INFORM ONDISC available from UMI/Data Courier: covers last 5 years, updated monthly. No printed version. Important for access to trade and scholarly literature on telecommunication business, economics, and management. Coverge more inclusive than WILSON BUSINESS ABSTRACTS (0492). Indexes about 800 scholarly and trade journals.

0491   **BUSINESS NEWS.** [Online service]. Columbus, OH: OCLC, 1992–date, updated daily. Database updated by Individual, Inc.'s HEADSUP service. "Brief summaries of news stories drawn from over 350 sources. ...Business News contains one to two weeks' worth of summaries and includes such subjects as information technology, telecommunications, health care, and defense. News summaries are added to the database daily. Once a week, summaries from the oldest week are deleted." Subject, source, and other indexing. Covers industry-monitoring newsletters COMMUNICATIONS DAILY (1037), COMMUNICATIONS WEEK (1089); wire services Reuters, PR Newswire; and selected major newspapers. Also covers press releases from the FCC. An alternative to similar services offered by LEXIS/NEXIS (0728), NewsNet, Dialog, and other vendors.

0492   **BUSINESS PERIODICALS INDEX.** New York: H. W. Wilson, 1958–date, monthly. Online version WILSON BUSINESS ABSTRACTS available from Wilsonline, OCLC: covers June 1982–date, updated twice weekly. CD-ROM version WILSON BUSINESS ABSTRACTS available from H. W. Wilson, Silver Platter: covers 1983–date, updated quarterly. Indexes telecommunication's trade and scholarly business, management, and economics literature. Covers over 300 English-language periodicals in advertising, broadcasting, communications, computer technologies and applications, industrial relations, international business, marketing, printing, publishing, public utilities, regulation of industry, and public relations. Particularly useful for identifying business literature on telecommunication technologies and organizations, including subject headings for facsimile transmission, electronic mail, local area networks, AT&T, Federal Communications Commission, International Telecommunications Union, and International Telecommunications Satellite Organization. Electronic versions offer abstracts and subject headings to improve access.

0493   **F & S INDEX UNITED STATES.** Foster City, CA: Information Access, 1968–date, monthly. Online version F & S INDEX PLUS TEXT available from Dialog, BRS, Data-Star: coverage varies, updated weekly. CD-ROM version available from Information Access: covers current year only, updated monthly. Formerly Predicasts' PREDICASTS F & S INDEX and variant titles. Covers over 1,000 financial, business, trade and industry journals and special reports. Part 1 indexes articles about industries by SIC codes; part 2 indexes articles about specific companies. Black dots identify major articles. Complemented by F & S INDEX INTERNATIONAL (1968–date) as well as PREDICASTS FORECASTS (1960–date) for statistics. Electronic versions combine printed indexes and feature full texts of 80% of articles.

0494   **INDEX OF ECONOMIC ARTICLES.** Pittsburgh: American Economics Association, 1886–date, annual. Variant titles. Online version ECONOMIC LITERATURE INDEX available on Dialog, Knowledge Index: covers 1969–date, updated quarterly. CD-ROM version ECONLIT available from Silver Platter: covers 1969–date, updated quarterly. Comprehensive bibliography covers 300 major economic journals, books, and collections.

4-A-3. Directories

0495   **BUSINESSORGS.** [Online service]. Columbus, OH: OCLC, 1993–date, updated annually. Corresponds to current editions of BUSINESS ORGANIZATIONS, AGENCIES, AND PUBLICATIONS DIRECTORY (Detroit: Gale, 1986–date, annual). Subtitle: "Directory of more than 26,000 new and established organizations, agencies, and publications worldwide." Subject, organization name, geographic (city), and other searching.

0496   **CORPORATE TECHNOLOGY DIRECTORY.** Wellesley Hills, MA: Corporate Technology Information Services, 1986–date, annual. Online version available from Orbit, Data-Star: covers current edition, updated quarterly. CD-ROM version available from Bowker Electronic Publishing in SCI-TECH REFERENCE PLUS: covers current edition, updated annually. Brief profiles of more than 35,000 high technology domestic and foreign companies arranged by "Corporate Technology Code" (under "Telecommunication" and many subdivisions). Company name, geographical, technology, product kind "who makes what" indexes. Glossary.

See 0382. **DIRECTORY OF AMERICAN RESEARCH AND TECHNOLOGY.**

0497   **DIRECTORY OF CORPORATE AFFILIATIONS.** Skokie, IL: National Register Publishing, 1973–date, annual. Online version CORPORATE AFFILIATIONS ONLINE available from Dialog: covers current edition, updated quarterly. CD-ROM version CORPORATE AFFILIATIONS ONDISC available from Dialog Ondisc. Acquisition and merger information for more than 5,000 major domestic corporations and 40,000 divisions, subsidiaries, and affiliates. Geographical, SIC Code, personal name indexes. Electronic versions cumulate DIRECTORY OF CORPORATE AFFILIATIONS, INTERNATIONAL DIRECTORY OF CORPORATE AFFILIATIONS, and DIRECTORY OF LEADING PRIVATE COMPANIES.

0498   **MILLION DOLLAR DIRECTORY.** Parsippany, NJ: Dun's Marketing Services, 1979–date, annual. Continues DUN AND BRADSTREET MILLION DOLLAR DIRECTORY and DUN AND BRADSTREET MIDDLE MARKET DIRECTORY (1959–1978). Online version available from Dialog: covers current edition: updated annually. CD-ROM version available from Dun's Marketing Services: covers current edition, udpated quarterly (also includes information from Dun's REFERENCE BOOK OF CORPORATE MANAGEMENT). Identifies 160,000 U.S. businesses and domestic subsidiaries of foreign companies with net worth of $500,000. Geographic and SIC Code indexes. Better electronic access to the same information and more is available in CD-ROM version DUN'S BUSINESS LOCATOR from Dun and Bradstreet Information Services: covers current edition; updated semiannually. Cumulating MILLION DOLLAR DIRECTORY, DUN'S BUSINESS RANKINGS, DUN'S INDUS-TRIAL GUIDE, DUN'S CONSULTANTS DIRECTORY (0583), DUN'S REGIONAL BUSINESS DIRECTORIES, DUN'S DIRECT ACCESS, and DUN'S EMPLOYMENT OPPORTUNITIES DIRECTORY, this gives data for 8.6 million U.S. companies in 15,000 categories.

0499   **ELECTRONICS MANUFACTURERS DIRECTORY: A MARKETERS GUIDE TO MANUFACTURERS IN THE UNITED STATES AND CANADA.** Twinsburg, OH: Harris, 1969–date, annual. Formerly U.S ELECTRONICS INDUSTRY DIRECTORY and WHO'S WHO IN ELECTRONICS. Basic information for companies: telephone and fax numbers, sales, top personnel, SIC Codes for products. Geographical and product (SIC Codes) indexes.

0500   **ENCYCLOPEDIA OF BUSINESS INFORMATION SOURCES.** Detroit: Gale, 1970–date, irregular. Subject listings for research resources (printed and electronic indexes, bibliographies, handbooks, directories, trade and professional organizations, research centers). Relevant materials under "Telecommunications," "Communications systems," "Communications satellites," "Cable television industry," "Computer communications," "Electronics industry," "Mobile telephone industry," "Public utilities," "Telephone industry," "Videotext/teletex." Formerly EXECUTIVE'S GUIDE TO INFORMATION SOURCES (1965).

0501   **INFORMATION INDUSTRY DIRECTORY.** Detroit: Gale, 1994 (14th ed.), 2 vols. "An international guide to organizations, systems, and services involved in the production and distribution of information in electronic form." Covers some 5,000 organizations in U.S and 70 countries. Indexes for corporate name and keyword, databases, publications, software, function or service, personal name, geographic, and subject. Formerly ENCYCLOPEDIA OF INFORMATION SYSTEMS AND SERVICES (1971–1990). Overlaps coverage of TELECOMMUNICATIONS DIRECTORY (0038).

See 0294 **INTERNATIONAL DIRECTORY OF COMPANY HISTORIES**.

0502   **NATIONAL TRADE AND PROFESSIONAL ASSOCIATIONS OF THE UNITED STATES.** Washington: Columbia Books, 1966–date, annual. 29th ed. (1994). Formerly NATIONAL TRADE AND PROFESSIONAL ASSOCIATIONS OF THE UNITED STATES AND CANADA AND LABOR UNIONS and variant titles. Gives directory data, describes activities, and lists publications for 7,500 active national trade associations, labor unions, professional scientific or technical societies, and other national organizations. Indexes for subjects, geographic, budget, executives, and acronyms. Relevant listings under "Communications," "Electricity & electronics," "Office equipment," "Radio-TV," "Standards," "Telephones," "Utilities."

0503   North American Telecommunications Association. **TELECOMMUNICATIONS SOURCE BOOK**. Washington: NATA, 1982–date, annual. Offers annual industry and policy overview; directory of contractors, manufacturers, suppliers, services, and users of equipment; descriptions of NATA membership structure and affiliate organizations.

See 0039 **RESEARCH CENTERS DIRECTORY**.

0504   **RESEARCH SERVICES DIRECTORY.** Detroit: Gale, 1981-date, irregular. Online version available on Dialog. 5th ed. (1993) describes over 4,400 U.S. commercial firms and labora-tories, including corporations, that conduct proprietary and non-proprietary research. Entries give brief data, including annual research budgets. Geographical, name, subject, and other indexes.

0505   **STANDARD AND POOR'S REGISTER OF CORPORATIONS, DIRECTORS AND EXECUTIVES.** New York: Standard and Poor's, 1928–date, annual. Supplements issued in April, July, and October. Online version (in several similarly names files) available from Dialog, LEXIS/NEXIS (0728): covers current edition; updated quarterly. CD-ROM version available in STANDARD AND POOR'S CORPORATIONS from Dialog Ondisc: covers current edition; updated quarterly. 67th ed. for 1994. Standard directory of U.S. companies. Data for 55,000 companies in vol. 1 and for over 500,000 executives in vol. 2. Vol. 3 contains SIC, geographic, and corporate indexes.

0506   **TELEPHONE ENGINEER & MANAGEMENT DIRECTORY.** Wheaton, IL; Harcourt, Brace, Jovanovich, 1936–date, annual. Variant titles. Directory and catalog of telephone companies and equipment.

0507   **TELEPHONE INDUSTRY DIRECTORY AND SOURCEBOOK.** Potomac, MD: Phillips Publishing, 1987–date, annual. Classified descriptions of 11,000 manufacturers, distributors, suppliers, carriers, regulatory agencies, consulting and marketing organizations, law firms, trade associations. Company, product, services indexes.

4-B. Statistics

4-B-1. Guides to and Indexes of Statistics

0508   **GUIDE TO U.S. GOVERNMENT STATISTICS.** McLean, VA: Documents Index, 1990–date, biennial. First published in 1956; with 1990–91 (5th) edition revised in alternate years. Arranges statistical sources by departments and agencies. Agency, world area, and U.S. area indexes.

0509   Kurian, George Thomas. **SOURCEBOOK OF GLOBAL STATISTICS.** London: Longman; New York: Facts on File, 1985, 413 pp. Describes 209 sources of statistical information; international in scope. Subject index.

0510   **PREDICASTS' BASEBOOK.** Foster City, CA: Predicasts, 1973–date, annual. Measures market size and cyclical sensitivity of products and industries. Arranged by SIC code with alphabetical list of products.

0511   **PREDICASTS' FORECASTS.** Foster City, CA: Information Access, 1960–date, quarterly. Formerly published by Predicasts. Short and long term statistics for economic indicators, products, markets, and industries, arranged by SIC code, extracted from trade and business press. Data cross references sources (journal title, date, page). Alphabetical list of products: "Electrical and electronic equipment" and "Communications" under SIC code 480. Coverage complements that provided in STATISTICAL MASTERFILE (0513).

0512   **STATE AND LOCAL STATISTICS SOURCES.** Gale, 1990–date, biennial. Classified listing of sources for each state. Telephone, radio, telecommunication, etc., sources under "Communications."

0513   **STATISTICAL MASTERFILE.** [CD-ROM.] Washington: Congressional Information Service, 1980–date, quarterly. CD-ROM indexing and abstracting service which cumulates CIS's printed AMERICAN STATISTICS INDEX (1973–date), INDEX OF INTERNATIONAL STATISTICS (1983–date), and STATISTICAL REFERENCE INDEX (1980–date). Online version only of AMERICAN STATISTICS INDEX available from Dialog: covers 1973–date, updated monthly. The CD-ROM version of ASI, IIS, and SRI is the most comprehensive and convenient index for statistics sources, covering the mid–1960s to the present. Detailed analyses of statistical data featured in wide range of publications issued by U.S. and international government and quasi-government agencies, trade and professional organizations, and in financial, business, and industry journals. Covers publications relevant to telecommunication issued by departments of Commerce, Labor, and Agriculture (Rural Telephone Program), FCC, NTIA, OTA; EC, OECD, ITU, foreign governments; and EIA, NATA, NARUC, state public utility commissions, research centers. Also identifies telecommunication statistics featured in weekly issues of TELEPHONY (1095) and ELECTRONIC BUSINESS, as well as journals like BUSINESS WEEK and FORBES, complementing coverage of PREDICASTS FORECASTS (0511). Full texts of selected statistical documents available in ASI, IIS, and SRI microfiche collections.

4-B-2. U.S Government Statistical Sources

0514   Department of Commerce. Bureau of the Census. **ANNUAL SURVEY OF COMMUNI-CATION SERVICES.** Washington: GPO, 1992-date, annual. Series begins with data for 1990 and covers telephone, broadcast, and cable industries (all part of the 48 series in the SIC code). A dozen charts and as many tables plus appendices on the research process behind the report are provided.

0515 _____. **COUNTY AND CITY DATA BOOK.** Washington: GPO, 1949–date, annual. CD-ROM version available from GPO. Data for states, regions, and divisions for 203 measured indicators.

0516 _____. **COUNTY BUSINESS PATTERNS.** Washington: GPO, 1946–date, annual. CD-ROM version available from GPO. Statistics on business activities throughout U.S.

0517 _____. **CURRENT CONSTRUCTION REPORTS: VALUE OF NEW CONSTRUCTION PUT IN PLACE.** Washington: GPO, 1959–date, monthly. Includes data on public telecommunications utilities and other construction. Indexed in STATISTICAL MASTERFILE (0513).

0518 _____. **CURRENT INDUSTRIAL REPORTS.** Washington: GPO, 1959–date, monthly. This series consists of more than 100 monthly, quarterly, and annual reports giving current data on different industries arranged by SIC codes. Series MA 36P "Communication Equipment, Including Telephone, Telegraph and other Electronic Systems and Equipment"; MA 26M "Radio and Television Receivers, Phonographs, and Related Equipment." Based on CENSUS OF MANUFAC-TURING. Indexed in STATISTICAL MASTERFILE (0513).

0519 _____. Bureau of Economic Analysis. **SURVEY OF CURRENT BUSINESS.** Washington: GPO, 1921–date, monthly. Formerly issued by Bureau of Census and other offices. Comprehensive report on economic conditions and business activities. Indexed in STATISTICAL MASTERFILE (0513).

See 0780. Federal Communications Commission. **ANNUAL REPORT.**

0520 Federal Communications Commission. **FIBER DEPLOYMENT UPDATE.** Washington: Data Communications Company, 1986–date, annual. Mileage of optical fiber lines put in place by long distance, RBOCs, and selected local carriers. April 1993 issue compiled by Jonathan M. Kraushaar. Indexed in STATISTICAL MASTERFILE (0513).

0521 _____. **STATISTICS OF COMMUNICATIONS COMMON CARRIERS.** Washington: FCC/GPO, 1937–date, annual. Telephone and telegraph companies financial and operating data (assets, liabilities, income, taxes, expenses, dividends, operating revenues, facilities, services, employment, and employee compensation), controlling companies, and Communications Satellite Corporation, based on company reports filed with FCC. Formerly STATISTICS OF THE COMMUNICATIONS INDUSTRY IN THE U.S. (1937–1957). Indexed in STATISTICAL MASTERFILE (0513).

0522 _____. **TELEPHONE RATES UPDATE.** Washington: Data Communications Company, 1988–date, semiannual. Compiled by James L. Lande. Report on local telephone rates, telephone service price indexes, and low-income subsidies, 1979–date. Data from Bureau of Labor Statistics, Common Carrier Bureau annual local rates surveys, and STATISTICS OF COMMUNICA-TIONS COMMON CARRIERS (0521). Indexed in STATISTICAL MASTERFILE (0513).

0523 _____. **TRENDS IN THE INTERNATIONAL COMMUNICATIONS INDUSTRY.** Washington: Data Communications Company, 1991–date, annual. Annual report on international telecommunications traffic and carrier revenues. June 1993 issue covers 1975–1991. Indexed in STATISTICAL MASTERFILE (0513).

0524 International Trade Commission. **U.S. INDUSTRIAL OUTLOOK.** Washington: GPO, 1961–date, annual. Trends and projections for over 350 manufacturing and nonmanufacturing industries, arranged by SIC code for "Information and Communications." Data on employment, worker earnings, capital expenditures, imports and exports, etc. Cumulates data from other Department of Commerce publications. Indexed in STATISTICAL MASTERFILE (0513).

0525   National Aeronautics and Space Administration. **SATELLITE SITUATION REPORT.** Greenbelt, MD: Goddard Space Flight Center, 1960–date, quarterly. Current title with 1979 volume. Inventory and data for all objects from all sources launched into orbit or deep space since Sputnik in 1957. Based on data provided by Goddard Space Research Laboratory, NORAD, or satellite owners. Indexed in STATISTICAL MASTERFILE (0513).

0526   Rural Telephone Bank, Rural Electrification Administration. **STATISTICAL REPORT, RURAL TELEPHONE BORROWERS.** Washington: GPO, 1939–date, annual. Data on rural telephone prorgam loans made by Rural Electrification Administration and Rural Telephone Bank: loans, operating, and financial statistics for borrowers. Indexed in STATISTICAL MASTERFILE (0513).

4-B-3. State Statistical Sources

0527   Illinois Commerce Commission. **OPERATING STATISTICS OF TELEPHONE COMPANIES IN ILLINOIS.** Springfield: Illinois Commerce Commission, 1979–date, annual. Financial and operating data for telephone companies and telephone cooperatives. Indexed in STATISTICAL MASTERFILE (0513).

0528   Mississippi State University, College of Business and Industry. **MISSISSIPPI STATIS-TICAL ABSTRACT.** Mississippi State, MS: Division of Business Research, college of Business and Industry, Mississippi State University, 1969–date, annual. Data on state telecommunications utilities. Indexed in STATISTICAL MASTERFILE (0513).

0529   New York State Department of Public Services, New York State Public Services Commission. **FINANCIAL STATISTICS OF THE MAJOR PRIVATELY OWNED UTILITIES IN NEW YORK STATE.** Albany: New York State Department of Public Services, ca 1973–date, annual. Indexed in STATISTICAL MASTERFILE (0513).

0530   Oregon Public Utility Commission. **OREGON UTILITY STATISTICS.** Salem: Oregon Public Utility Commission, 1983–date, annual. Indexed in STATISTICAL MASTERFILE (0513).

0531   Washington State Utilities and Transportation Commission. **STATISTICS OF UTILITY COMPANIES, WASHINGTON STATE.** Olympia: Washington State Utilities and Transportation Commission, ca 1948–date, annual. Indexed in STATISTICAL MASTERFILE (0513).

0532   Wisconsin Public Service Commission. **STATISTICS OF WISCONSIN PUBLIC UTILITIES.** Madison: Wisconsin Public Service Commission, 1940–date, annual. Indexed in STATISTICAL MASTERFILE (0513).

4-B-4. Private Statistical Sources

0533   AT&T. **BELL SYSTEM STATISTICAL MANUAL.** New York: AT&T Comtroller's Accounting Division, 19xx-1983, annual. Detailed statistics from 1950 to date of publication on all aspects of AT&T service, equipment usage, and finances. Designed for internal use only, this is invaluable.

0534   _____. **SELECTED INTERSTATE DATA UNDER THE DIVISION OF REVENUE CONTRACTS AND LONG LINES STATISTICS.** New York: AT&T, 1960-1983, annual. Detailed AT&T indicators describing long distance services, and revenue divisions with local Bell operating companies.

0535   _____. **STATISTICAL REPORT.** New York: AT&T, annual published through 1983. Supplement to the firm's regular annual report, this provides brief summary statistics on the central company and its major subsidiaries.

0536   Electronic Industries Association. **CONSUMER ELECTRONICS: U.S. SALES.** Washington: Electronic Industries Association, 1984–date, semiannual. Current and historical unit and dollar sales for product categories (audio, video, home information, accessories, etc.) and total factory sales for industry; trends and household penetration rates. Indexed in STATISTICAL MASTERFILE (0513).

0537 _____. **ELECTRONIC MARKET DATABOOK.** Washington: Electronic Industries Association, 1969–date, annual. Yearbook of U.S. electronics industry giving sales and production facts and figures for consumer electronics, telecommunications and defense-related communications equipment, computers and peripherals, government electronics, and international trade. Tables, graphs, charts. Glossary. Data largely based on U.S. departments of Commerce and Labor reports, compiled and synthesized by EIA Marketing Services. Formerly ELECTRONIC INDUSTRIES YEARBOOK and ELECTRONIC INDUSTRIES REVIEW (1969). Indexed in STATISTICAL MASTERFILE (0513).

0538 _____. **ELECTRONIC MARKET TRENDS.** Washington: Electronic Industries Association, 1984–date, monthly. Economic information and analysis for U.S. electronics industry: shipments, trade, employment, imports and exports, by economic indicators, producer and consumer price indexes. Based on U.S. Commerce Department data. Indexed in STATISTICAL MASTERFILE (0513).

0539 _____. **U.S. CONSUMER ELECTRONICS INDUSTRY IN REVIEW.** Washington: Electronic Industries Association, 1968-date, annual. Title has varied. Issue for 1994 is 110 pp. of text, tables, and charts covering data from the late 1980s into the early 1990s. Details units sold and factory value for a wide variety of consumer electronics. Indexed in STATISTICAL MASTERFILE (0513).

0540 **INVESTEXT.** [Online service]. Boston: Thomson Financial Services, 1982–date, daily. Available from Data-Star, Dialog, LEXIS/NEXIS (0728), NewsNet. CD-ROM version available from Thomson. Online full-texts of investment analyses and reports on companies and industries by more than 300 leading investment banks, consulting and research firms, and other financial authorities. Global coverage.

0541 North American Telecommunications Association. **TELECOMMUNICATIONS MARKET REVIEW AND FORECAST.** Washington: NATA, 1990–date, annual. Data on telecommunications equipment and service industries market trends and outlook (finances, operations, foreign trade); arranged by markets (equipment manufacturing and services, local exchanges, interexchanges and networks, customer premises products, mobile communications and computer-telephone integration). Selected company data. Data compiled by NATA Market Research. Indexed in STATIS-TICAL MASTERFILE (0513).

0542 **RMA ANNUAL STATEMENT STUDIES.** Philadelphia: Robert Morris Associates, 1923–date, annual. 71st annual edition (1993). Composite financial data for manufacturing, wholesaling, retailing, service, and contracting industries: ratios for liquidity, coverage, leverage, operating, expenses to sales. Useful for comparing companies' performances. Bibliography. SIC code and subject indexes. Formerly STATEMENT STUDIES (1923–1962) and ANNUAL STATEMENT STUDIES (1964–1976). Indexed in STATISTICAL MASTERFILE (0513).

0543 Rubin, Michael Rogers, Mary Taylor Huber, with Elizabeth Lloyd Taylor. **THE KNOWLEDGE INDUSTRY IN THE UNITED STATES, 1960–1980.** Princeton, NJ: Princeton University Press, 1986, 213pp. Statistics on communications and information services in U.S. with narrative and tables on education, research and development, communication media, information machines, information services. Tables, bibliographical references, index.

0544 **STANDARD AND POOR'S INDUSTRY SURVEYS.** New York: Standard and Poor's, 1959–date, quarterly. Basic data on 69 domestic industries, including electronics and telecommunications. Information on prospects, trends, problems, with statistics composite industry data, and financial comparisons of leading companies. Kept up-to-date by weekly loose-leaf service.

See 1004. **THE WORLD'S TELEPHONES.**

4-C. Annual Reports and 10-K Forms

0545   **COMPACT D/SEC.** [CD-ROM]. Bethesda, MD: Disclosure Incorporated, 1985– date, monthly. Online version DISCLOSURE available from LEXIS/NEXIS (0728), OCLC (DISCLOSURE/WORLDSCOPE SNAPSHOTS): covers 1982–date, updated weekly. Information on more than 12,000 public companies extracted from 10Ks, 20Fs, 10Qs, 8Ks, annual reports, tender offer and acquisitions reports, proxy statements, etc. No printed version. Information includes 2-year comparisons of balance sheets, 3-year comparisons of income statements, 5-year summary of operating income, sales and earnings per share.

0546   **FAIRCHILD'S ELECTRONIC INDUSTRY FINANCIAL DIRECTORY.** New York: Fairchild Publications, 1962–date, annual. Formerly ELECTRONIC NEWS FINANCIAL FACT BOOK AND DIRECTORY (1962–1991). Covers about 1,000 companies. Concise profiles detail products and services, subsidiaries, acquisitions, transfer agents, employees, sales and earnings, common stock, income accounts, assets, liabilities, etc.

0547   **MOODY'S MANUALS.** New York: Moody's Investors Services, 1909–date, annual. Online updating for 13,000 companies available in MOODY'S CORPORATE NEWS—U.S. from Dialog: updated weekly. CD-ROM versions available from Moody's Investor Service include MOODY'S INDUSTRIAL DISC, MOODY'S INTERNATIONAL COMPANY DATA, MOODY'S PUBLIC UTILITY DISC: covers current editions; updated quarterly. Contain information on publicly owned companies in particular sectors, including income statements, balance sheets, financial and operating ratios based on company reports and SEC filings; historical and current descriptions with lists of subsidiaries; details on capital structure, capital stock and long term debt. Each manual includes geographical and business classification indexes (blue pages) and updated by looseleaf NEWS REPORTS. Complete historical series of manuals available on microfiche, MOODY'S MANUALS ON MICROFICHE. Most useful volumes for telecommunications include:

　　　　1 **MOODY'S INDUSTRIAL MANUAL** (1909–date). Covers about 2,000 companies.

　　　　2 **MOODY'S INTERNATIONAL MANUAL** (1981–date). Covers about 7,000 companies in 90 countries, excluding U.S.

　　　　3 **MOODY'S PUBLIC UTILITY MANUAL** (1914–date). Covers all public and most privately held utilities.

0548   **STANDARD & POOR'S CORPORATION RECORDS.** New York: Standard and Poor's, 1940–date, bimonthly. Updated by STANDARD AND POOR'S DAILY CORPORATION NEWS. Online version corresponds in part to STANDARD AND POOR'S CORPORATE DESCRIPTIONS from LEXIS/NEXIS (0728): covers current edition; updated biweekly; and in STANDARD AND POOR'S CORPORATE DESCRIPTIONS PLUS NEWS from Dialog, Knowledge Index: covers current edition; updated biweekly. CD-ROM version corresponds in part to STANDARD AND POOR'S CORPORA-TIONS from Dialog Ondisc: covers current edition; updated monthly. Information on publicly owned companies: capitalization, corporate backgroud, stock data, bond descriptions, earnings, and finance. Useful features include CEO annual messages to stockholders and management discussions.

4-D. Employment Data

0549   Bureau of Labor Statistics, U.S. Department of Labor. **BULLETIN OF THE UNITED STATES BUREAU OF LABOR STATISTICS.** Washington: GPO, 1913–date, irregular. A complex series of reports that publishes a wealth of information relevant to telecommunications and infor-mation technologies, especially their impact on different industries (including their own). Generally reports on employment trends, labor conditions, etc. in different industrial sectors. Series (1915–1959) indexed in BULLETIN 1281. Formerly BULLETIN OF THE BUREAU OF LABOR. Various numbers relevant to telecommunications include:

　　　　1 **OCCUPATIONAL EMPLOYMENT IN SELECTED NONMANUFACTURING INDUSTRIES. BULLETIN, 2417** (1991). Covers communications industries.

2 **OUTLOOK FOR TECHNOLOGY AND LABOR IN TELEPHONE COMMUNICA-TIONS. BULLETIN, 2357** (1990). "Appraises some of the major technological changes emerging in' telephone communications (SIC 481) and discusses the impact of these changes on productivity and labor over the next 5 to 10 years" (p. iii). Charts, statistics, bibliography of other BLS publications on technology's impact on industries.

3 **INDUSTRY WAGE SURVEY: COMPUTER AND DATA PROCESSING SERVICES. BULLETIN, 2318** (1988); also **BULLETIN, 2028** (1979); **BULLETIN, 1909** (1976).

4 **OCCUPATIONAL EMPLOYMENT IN TRANSPORTATION, COMMUNICA-TIONS, UTILITIES, AND TRADE. BULLETIN, 2220** (1985).

5 **INDUSTRY WAGE SURVEY: COMMUNICATIONS. BULLETIN, 2100** (1981); also **BULLETIN, 1991** (1976).

0550 _____. **OCCUPATIONAL OUTLOOK HANDBOOK.** Washington: GPO, 1949– date, annual. (BULLETIN OF THE UNITED STATES BUREAU OF LABOR STATISTICS, number varies). CD-ROM version available from GPO. A standard reference source giving data for employment trends by occupation. Describes occupations and gives basic employment data for job outlook and trends, working conditions, training and qualifications, earnings, as well as identifying related occupations and sources of additional information. Entries arranged by "DOT" (Dictionary of Occupational Titles) class numbers, with cross referenced "SOC" (Standard Occupational Classification) numbers; for example, DOT 236.562–010 for "Telegrapher," DOT 722.131–010 for "Communications-equipment supervisor." Subject index offers common access. Chapters on communications occupations have been reprinted as separate BLS monographs; for example, COMMUNICATIONS, DESIGN, PERFORMING ARTS, AND RELATED OCCUPATIONS (1984).

0551 _____. Office of Employment and Unemployment Statistics. **OCCUPATIONAL EMPLOYMENT SURVEY BOOKLET OF DEFINITIONS: COMMUNICATIONS.** Washington: Bureau of Labor Statistics, 1990, 14 pp. Occupational definitions for clerical, skilled, and professional positions in communications industry.

0552 National Science Foundation. National Science Board. **SCIENCE & ENGINEERING INDICATORS.** Washington: GPO, 1972–date, biennial. Title has varied. Overviews and data for science and technical education and research in the U.S. Chapters cover all aspects of elementary, secondary, and higher education, and postgraduate education; science and engineering workforce in by industry sector and R&D; federal and industry funding and support for R&D; academic R&D; technology development, patents and inventions, markets, etc.; public attitudes toward science and technology. Discussions include international comparisons. Bibliographies, index. Useful references under "Electrical/electronics engineering" and other headings.

0553 _____. **SCIENTISTS, ENGINEERS, AND TECHNICIANS IN NONMANUFAC-TURING INDUSTRIES.** Washington: National Science Foundation, 1987–date, triennial. Employment estimates by industry and occupation. Data based on Bureau of Labor Statistics and state employment security services' Occupational Employment Statistics surveys. Indexed in STATIS-TICAL MASTERFILE (0513).

0554 _____. **SCIENTISTS, ENGINEERS, AND TECHNICIANS IN TRADE AND REGULATED INDUSTRIES.** Washington: National Science Foundation, 1988–date, triennial. Latest edition covering 1991 data published in 1993. Report covers professional employment in communications and public utilities industries as well as in other retail and wholesale trades, transportation. Indexed in STATISTICAL MASTERFILE (0513).

4-E. Selected Associations and Organizations

4-E-1. Trade Associations

0555 **Alliance for Telecommunications Industry Solutions.** 1200 G Street NW, Washington, DC 20005. Voice: 202 628 6380. Fax: 202 393 5453. Formerly Exchange Carriers Standards Association (1983–1993). Members include domestic wireline exchange carriers (about 150 companies). Serves as forum for local exchange interchange carriers, manufacturers, vendors, end users, cellular service providers, competitive access providers, cable television services. Major telecommunications standards developing organization. ATIS/ECSA's Standards Advisory Committee (SAC) represents T1 in ANSI and sponsors Standards Committee T1, comprised of exchange carriers, interexchange carriers and resellers, manufacturers and vendors, and general interests (government, user groups, consultants, committee liaisons), that develops network interconnect standards in technical subcommittees: Network Interfaces (TIE1); Internetwork Operations, Administration, Maintenance (OAM), and Provisioning (T1M1); Performance (T1Q1); Services, Architecture, and Signaling (T1S1); Digital Hierarchy and Synchronization (T1X1); and Specialized Subjects (T1Y1). ATIS/ECSA's Exchange Telephone Group Committee represents exchange carriers in ANSI. Publishes ATIS NEWS (1983–date, bimonthly; ATIS ANNUAL REPORT (1983–date, annual); TELECOMMUNICATIONS COMMITTEE T1 (1983–date, semiannual); T1 COMMITTEE ANNUAL REPORT (1983–date, annual). Also publishes many technical working papers that assess existing and developing standards.

0556 **American Electronics Association.** 5201 Great American Pkwy., Suite 520, Santa Clara, CA 95054. Voice: 408 987 4200. Fax: 408 970 8565. Formerly Western Electronic Manufacturers Association (1943-1971). Represents 3,400 companies in electronics and information technologies industries. Participates in standards development. Maintains Washington DC and regional offices. Annual budget of about $16.8 million. Sponsors American Electronics Association Political Action Committee. Publishes monographs, like SETTING THE STANDARD: A HANDBOOK ON SKILL STANDARDS FOR THE HIGH TECH INDUSTRY (1994); SOURCES OF ELECTRONICS INDUSTRY INFORMATION IN JAPAN (1985); and AEA ENGLISH-JAPANESE, JAPANESE-ENGLISH ENGINEERING TERMINOLOGY DICTIONARY, 2nd ed. (1988). Also publishes proceedings (1958-date), industry surveys, and membership directory.

See 0399. **American National Standards Institute**.

0557 **Cellular Telecommunications Industry Association**. 1133 21st St. NW, 3rd Floor, Washington DC 20036-3390. Voice: 202 785 0081. Fax: 202 785 0721. Formerly Cellular Communications Industry Association (1984); absorbed Cellular Radio Communications Association (1985). Major cellular telecommunications organization. Members include nearly 400 companies (about 90% of industry) holding cellular licenses, permits, or authorizations from FCC, including system operators, equipment manufacturers, engineering firms, and others in cellular telephone and mobile communications industry. Annual budget of about $10 million. Represents industry in legislation and regulation. Publishes monographs, like BRINGING INFORMATION TO PEOPLE: CELEBRATING THE WIRELESS DECADE (1993); WIRELESS SOURCEBOOK (1993-date, annual); STATE OF THE CELLULAR INDUSTRY (1991-date, annual); and other reports.

0558 **Computer and Business Equipment Manufacturers Association**. 1250 I St. NW, Washington DC 20005. Voice: 202 737 8888. Fax: 202 638 4922. Members include 28 companies (manufacturers, assemblers, and producers of information processing, business and communications products, supplies, and services). Participates in telecommunications standards development as Secretariat for the ANSI Committee on Information Processing Systems (X3). Monitors legislation, regulation, and issues related to international and domestic industry. Formerly Office Equipment Manufacturers Institute (1916–1962) and Business Equipment Manufacturers Association (1963–1973). Semiannual meetings. Publishes CBEMA COMMENT (1979-date); INFORMATION TECHNOLOGY INDUSTRY GLOBAL MARKET ANALYSES (1989-date); regional reports, and annual report.

0559   **Electronic Industries Association** (EIA). Standards Sales, 2500 Wilson Blvd., Arlington, VA 22201. Voice: 703 907 7500. Fax: 703 907 7501. Fee-based research service: EIA Research Center. Voice: 703 907 7751. Major trade association of U.S. electronics manufacturers, representing 1,000 members. Annual budget of about $42 million. Monitors legislation and regulation and represents industry interests in competitiveness, common distribution, trade, and government procurement. Responsible for standards-developing in data communications. Represents industry in ANSI and ISO. EIA's telecommunication sector merged with US Telecommunication Suppliers Association (USTSA) in 1988 to form Telecommunication Industry Association (TIA); EIA Standards Department develops EIA/TIA standards in technical committees: Mobile Radio (TR-80); Microwave Radio (TR-14); Facsimile Systems and Equipment (TR-29); Data Transmission Systems and Equipment (TR-30); Personal Radio (TR-32); Satellite Equipment and Systems (TR-34); Telephone Terminals (TR-41); Cellular Radio (TR-45); Optical Communications (FO-2); Fiber Optics (FO-6). Sponsors Joint Electron Device Engineering Council, Electronics Political Action Committee, Electronic Industries Foundation, and Consumer Electronics Shows. Major statistical publications include ELECTRONIC MARKET DATA BOOK (0537); ELECTRONIC MARKET TRENDS (0538); ELECTRONIC FOREIGN TRADE (0999); and CONSUMER ELECTRONICS: U.S. SALES (0536). Also publishes INTERFACE (1984-date); proceedings; trade directories; research reports; and annual report. Semiannual meetings. Formerly Radio Manufacturers Association (1924-1950); Radio-Television Manufacturers Association (1950-1953); Radio-Electronics-Television Manufacturers Association (1953-1957). Includes Magnetic Recording Industry Association; absorbed Association of Electronic Manufacturers, Institute of High Fidelity, and Mobile Electronics Association.

0560   **Information Industry Association**. 555 New Jersey Ave. NW, Washington, DC 20001. Voice: 202 639 8262. Fax: 202 638 4403. Members include 500 companies that provide information services, products, and technologies. Divisions for databases and publishing, electronic services, financial services, and voice information services. Includes Directory Publishers Alliance. Publishes monographs, like Ronald L. Plesser's SERVING CITIZENS IN THE INFORMATION AGE: ACCESS PRINCIPLES FOR STATE AND LOCAL GOVERNMENT INFORMATION (1993); INFORMATION SOURCES (1983-date); and membership telephone directory.

0561   **International Communications Industries Association**. 3150 Spring St., Fairfax VA 22031-2399. Voice: 703 273 7200. Fax: 703 278 8082. 1,200 members, including video, audiovisual, and computer hardware and software dealers, manufacturers, and users. Interests in competition and regulation, legislation, small business issues, postal rates, copyright, education, and taxation. Participates in standards development. Sponsors Audio-Visual Communications Fund Political Action Committee. Operates Educational Communications Foundation. Headquarters for INFOCOMM International, annual trade show. Publishes COMMUNICATIONS INDUSTRY REPORT (1983-date, monthly); DIRECTORY OF VIDEO, COMPUTER, AND AUDIO-VISUAL PRODUCTS (1953-date), formerly AUDIO VISUAL EQUIPMENT DIRECTORY and variant titles; industry studies, including AUDIO-VISUAL EQUIPMENT SURVEY (1989-date); and other resources, like FOUNDATION GRANTS GUIDE: GRANTS WITH A SLICE FOR COMMUNICATIONS TECHNOLOGY PRODUCTS (1984). Holds annual meetings. Absorbed Association of Media Producers and International Media Producers Association. Formerly National Audio-Visual Association (1939-1983).

0562   **National Association of Broadcasters**. 1771 N St. NW, Washington, DC 20036-2891. Voice: 202 429 5300. Fax: 202 429 5343. Oldest (established 1922) and most important trade association of radio and television stations, and associates, including mass communication educators and academic programs. 7,500 members, with annual budget of about $18 million. Monitors legislation and regulation and represents industry interests of broadcasting. Sponsors Television and Radio Political Action Committee (TARPAC). Maintains a vigorous publication program; publishes reference works and handbooks, including Susan M. Hill's BROADCASTING BIBLIOGRAPHY: A GUIDE TO THE LITERATURE OF RADIO AND TELEVISION, 3rd ed. (1989) and LEGAL GUIDE TO BROADCAST LAW AND REGULATION (1994); and research studies, including MANY ROADS

TO HOME: THE NEW ELECTRONIC PATHWAYS (1988); Kenneth R. Donow and Suzanne G. Douglas' THE POTENTIAL IMPACT OF TELEPHONE REGIONAL HOLDING COMPANIES' DIVERSIFICATION AND VIDEO SERVICE STRATEGIES ON THE BROADCASTING INDUSTRY (1990); MULTIMEDIA 2000: MARKET DEVELOPMENTS, MEDIA BUSINESS IMPACTS AND FUTURE TRENDS (1993); Mark R. Dodd and Marcia L. De Sonne's SATELLITES AND RADIO BROADCASTING: HISTORICAL REVIEW AND MARKET DEVELOPMENTS (1991); TELCO FIBER AND VIDEO MARKET ENTRY: ISSUES AND PERSPECTIVES FOR THE FUTURE (1989); and Marcia L. De Sonne's SPECTRUM OF NEW BROADCAST/MEDIA TECHNOLOGIES (1989), among others. NAB also publishes proceedings of conferences and newsletters aimed at the radio, television, and technical segments of its membership. Merged with Television Broadcasters Association in 1951; absorbed Daytime Broadcasters Association in 1985, and National Radio Broadcasters Association in 1986. Formerly National Association of Radio and Television Broadcasters (1951-1957).

0563　**National Cable Television Association.** 1724 Massachusetts Ave. NW, Washington DC 20036. Voice: 202 775 3550. Fax: 202 775 3604. Members include 2,700 cable TV systems and 400 associated manufacturers, distributors, hardware suppliers, programmers, and other services. Annual budget of $29.7 million. Sponsors Cable Television Political Action Committee (Cable-PAC). Publishes reference works aimed at membership, like CABLE PRIMER (1984) and COMPLETE CABLE BOOK (1994); also publishes industry reports and studies, including AMERICA'S STUDENTS: TRAVELERS ON CABLE'S INFORMATION HIGHWAYS (1993) and THE NEVER-ENDING STORY: TELEPHONE COMPANY ANTICOMPETITIVE BEHAVIOR SINCE THE BREAKUP OF AT&T (1991). Also publishes conference proceedings and membership newsletters. Formerly National Community Television Association (1952-1967).

0564　**National Telephone Cooperative Association.** 2626 Pennsylvania Ave. NW, Washington, DC 20037-1695. Voice: 202 298 2300. Fax: 202 298 2320. Rural member-owned telephone companies and small independent commercial companies. Annual budget of about $11 million. Monitors legislation and regulation and assists members. Co-sponsors conferences that address rural technology issues. Publishes membership newsletters RURAL TELECOMMUNICATIONS (1093) and NCTA EXCHANGE (1993-date, bimonthly); industry surveys, including COMPENSATION & BENEFITS IN THE INDEPENDENT TELEPHONE INDUSTRY (1993-date, annual); annual reports; and membership directory. NTCA: OUR FIRST TWENTY-FIVE YEARS (1979) is organizational history.

0565　**North American Telecommunications Association.** 2000 M St. NW, Suite 550, Washington, DC 20036-3367. Voice: 202 296 9800. Fax: 202 296 4993. Toll free: 800 538 6282. Major trade organization of telecommunications manufacturers, suppliers, installers, and system users. Monitors legislation and regulation and serves as information and education clearinghouse and advocate for members. Sponsors NATA Political Action Committee. Supports large publication program: publishes industry's major trade journal, TELECOMMUNICATIONS (1039) and other membership newsletters and directories; reference works and statistical cumulations, including TELECOMMUNICATIONS SOURCEBOOK (0503), DIRECTORY OF TELECOMMUNICA-TIONS SCHOOLS AND INSTITUTIONS (0683), TELECOMMUNICATIONS MARKET REVIEW AND FORECAST (1990-date, annual), and TELECOMMUNICATIONS EXPORT GUIDE, 3rd ed. (1993); and industry reports and studies, like VOICE MESSAGING INDUSTRY REVIEW (1991-date, annual), REMOTE ACCESS TOLL FRAUD: DETECTION AND PROTECTION (1992); INDUSTRY BASICS: AN INTRODUCTION TO THE HISTORY, STRUCTURE, AND TECHNOLOGY OF THE TELECOMMUNICATIONS INDUSTRY (1991); and THE POST-DIVESTITURE U.S. TELECOMMUNICATIONS EQUIPMENT MANUFAC-TURING INDUSTRY: THE BENEFITS OF COMPETITION (1991). Formerly North American Telephone Association (1970-1982).

0566　**Software Publishers Association.** 1730 M St. NW, Suite 700, Washington, DC 20036-4510. Voice: 202 452 1600. Fax: 202 223 8756. Publishers of microcomputer software. Annual budget of about $9 million. Monitors legislation and regulation. Interests in software copyrights. Publishes membership newsletter INDUSTRY UPDATE (1989-date) and directory; reference works like SPA INTERNATIONAL RESOURCE DIRECTORY AND GUIDE, 13th ed. (1993) and Thomas J. Smedinghoff's THE SOFTWARE PUBLISHERS ASSOCIATION LEGAL GUIDE TO MULTI-MEDIA (1994); and research reports, including REPORT ON THE EFFECTIVENESS OF TECHNOLOGY IN SCHOOLS, 1990-1994 (1994).

0567   **Telecommunications Industry Association**. 2001 Pennsylvania Ave. NW, Suite 800, Washington, DC 20006-1813. Voice: 202 457 4912; 202 457 4912. Fax: 202 457 4939. Major trade organization for telecommunications services and equipment manufacturers and suppliers. Annual budget of about $7 million. Participates in standards development. Affiliated with EIA (0559) and United States Telephone Association (0569); co-publishes reports with EIA. Publishes membership newsletter, conference proceedings, and TIA DIRECTORY AND BUYERS GUIDE (1994). Formerly United States Telecommunications Suppliers Association (1988).

0568   **Telocator: The Personal Communications Industry Association**. 1019 19th St. NW, 11th Floor, Washington, DC 20036-5105. Voice: 202 467 4770. Fax: 202 467 6987. Members include FCC-licensed radio and cellular common carriers who provide public mobile communication services (paging, cellular radio telephones, PCS, and new wireless technologies). Monitors legislation and regulation regarding standards and policies and informs membership. Publishes PCIA JOURNAL (1092), formerly TELOCATOR, and membership directory. Formerly National Mobile Radio System, National Association of Radiotelephone Systems, and Telocator Network of America (1946-1987).

0569   **United States Telephone Association**. 1401 H St. NW, Washington DC 20005. Voice: 202 836 7300. Fax: 202 835 3187. Local exchange telephone companies and their manufacturers and suppliers. Annual budget of about $12.4 million. Provides information and training. Participates in FCC rulemaking. Sponsors USTA Political Action Committee. Holds annual meetings and tradeshows. Publishes membership newsletter TELETIMES (1990-date, quarterly); conference proceedings (1914-date); statistical reports, including STATISTICS OF THE LOCAL EXCHANGE CARRIERS (1989-date, annual), formerly STATISTICS OF THE INDEPENDENT TELEPHONE INDUSTRY; reference works, like PHONE FACTS (1984–date, annual); technical and economic reports, including CHANGING TECHNOLOGY AND TELECOMMUNICATIONS ASSETS (1990) and COST AND PRICING PRINCIPLES FOR TELECOMMUNICATIONS (1990); and other works, like THE HISTORY OF THE TELEPHONE INDUSTRY (1991) and THE RING OF SUCCESS: SEVENTY-FIVE YEARS OF THE INDEPENDENT TELEPHONE MOVEMENT (1972). Formerly Independent Telephone Association of America and National Independent Telephone (1897–1982).

0570   **Utilities Telecommunications Council**. 1140 Connecticut Ave. NW, Suite 1140, Washington, DC 20036. Voice: 202 872 0030. Fax: 202 872 1331. Interest in development and improvement of telecommunications used by electric, gas, and water utilities. Members include utility companies eligible to hold FCC Power Radio Service licenses. Participates in FCC rulemaking. Monitors legislation and regulation related to radio spectrum for fixed and mobile communication and telecommunications operations of energy utilities. Publishes UTC REPORTS (1985–date, monthly) and research studies, including COMPILATION OF COMMUNICATIONS STUDIES DEMON-STRATING POTENTIAL UTILITY INVOLVEMENT IN THE DEVELOPMENT OF THE NATIONAL INFORMATION INFRASTRUCTURE (1994). Annual meetings.

4-E-2. *Professional Organizations*

0571   **Association for Computing Machinery**. 1515 Broadway, New York, NY 10036. Voice: 212 869 7440. Fax: 212 869 0481. Internet: acmhelp@acm.org.85,000 individual members. Annual budget of $26.1 million. Sponsors conferences. ACM's SIGCOMM special interest group is a forum for discussion and research on computer networking's social, regulatory, and technology transfer issues. Publishes COMPUTER COMMUNICATIONS REVIEWS (1047); ACM TRANSACTIONS ON INFORMATION SYSTEMS (1989–date, quarterly); and other scholarly research journals. Also publishes conference proceedings and monographs. "ACM Press History Series" (with Addison-Wesley Publishers) includes A HISTORY OF MEDICAL INFORMATICS (1990) and A HISTORY OF PERSONAL WORKSTATIONS (1988).

0572   **Communication Workers of America.** 501 3rd St. NW, Washington, DC 20001-2797. Voice: 202 434 1100. Fax: 202 434 1279. Labor organization of 650,000 telecommunications workers. Annual budget of about $65 million. Sponsors CWA-COPE Political Action Committee. Publishes national newletter CWA NEWS (1941-date, monthly) and convention proceedings. Individual locals also publish newsletters. Also published CWA AT FIFTY: A PICTORIAL HISTORY OF THE COMMUNICATION WORKERS OF AMERICA, 1938-1988 (1988). Maintains archives (0228). Formerly National Federation of Telephone Workers (1938-1947). Merged with Telephone Workers Organizing Committee and absorbed the International Typographical Union in 1987 and United Telegraph Workers in 1990.

0573   **EDUCOM.** 1112 16th St. NW, Suite 600, Washington, DC 20036-4823. Voice: 202 872 4200. Fax: 202 872 4318. Internet: inquiry@educom.edu.Non-profit organization of colleges, universities, and other institutions to promote the use of information technology. Annual budget of $6 million. Publishes EDUCOM REVIEW (1989-date, quarterly) and EUIT NEWSLETTER (1990-date, monthly). Proceedings of annual meetings and other publications often published in audio or video, including ANNUAL REVIEW OF CAMPUS COMPUTING TECHNOLOGY, ACADEMIC AND REGULATORY TELECOMMUNICATIONS ISSUES (1991), and THE CAMPUSWIDE INFORMATION SYSTEM (1991). Formerly Interuniversity Communications Council (1964-1984).

0574   **Federal Communications Bar Association.** 1150 Connecticut Ave. NW, Suite 1050, Washington, DC 20036-4104. Voice: 202 833 2684. Fax: 202 833 1308. Attorneys in communications law who practice before the FCC, courts, and state and local regulatory agencies, nonattorneys, and law students. Publishes FEDERAL COMMUNICATIONS LAW JOURNAL (1113) and a membership newsletter. Also publishes many case and practice manuals and reference works, including INTERNATIONAL COMMUNICATIONS PRACTICE HANDBOOK (1993) and ENVIRONMENTAL CONCERNS FOR COMMUNICATIONS ENTITIES (1991), as well as a membership directory.

0575   **Institute of Electrical and Electronics Engineers** (IEEE). 345 East 47th St., New York, NY 10017-2394. Voice: 212 705 7900. Fax: 212 752 4929. Maintains IEEE Service Center: 445 Hoes Ln., P.O. Box 1331, Piscataway NJ 08855, voice: 908 981 0060; and Washington office: 1828 L. St. NW, Suite 1202, Washington, DC 20036, voice: 202 785 0017, fax: 202 785 0835. International technological and professional organization. 320,000 members in 130 countries. Promotes advancement of electronic technology and standards. Advises and comments by concensus on standards; identifies areas in need of standards and formally and systematically develops and publishes standards. IEEE Communications Society (COMSOC)'s technical committees may develop standards, coordinated by its Standards Liaison and Coordinating Committee. IEEE also supplies standards. Voice: 415 259 5040, 800 949 IEEE. Fax: 415 259 5045. Internet: askiee@ieee.org. ASKIEEE also on Dialog Dialmail. Sponsors numerous associations, societies, committees, and sections, including Aerospace and Electronics Systems, Antennas Propagation, Broadcast Technology, Communications, Geoscience and Remote Sensing, Industrial Electronics, Medical Imaging Committee, Robotics and Automation, among others. Publishes IEEE SPECTRUM (1053) and many other journals, proceedings, and publications. Maintains separate archives (0233). Formered by merger of American Institute of Electrical Engineers (1884–1964) and Institute of Radio Engineers (1912–1964).

4-E-3. Public Interest Groups

0576   **Alliance for Public Technology.** 901 15th St. NW, Washington, DC 20005. Voice: 202 408 1403. Fax: 202 408 1134. Internet: apt.apt.org. Also World-Wide Web: http://apt.org/apt.html. Public interest group with interest in affordable access to information and telecommunication services and technologies, especially for elderly, residential consumers, low-income groups, and disabled. Publishes newsletter APT NEWS (1991-date, bimonthly) and reports. Co-sponsored Telecommunity

Conference (1993) and report BRINGING HOME THE ELECTRONIC HIGHWAY: PUBLIC TV AND UNIVERSAL ACCESS (1994), by Susan G. Hadden.

0577   **Coalition for Networked Information.** c/o Association of Research Libraries, 1527 New Hampshire Ave. NW, Washington, DC 20036. Voice: 202 232 2466. Fax: 202 462 7849. Internet: joan@cni.org. Promotes "the creation of and access to information resources in networked environments in order to enrich scholarship and to enhance intellectual productivity." Major interest in library cooperation in electronic and networked environment. Publishes newsletter NEWS AND VIEWS (1991-date); sponsors reports and other publications, including HUMANITIES AND ARTS ON INFORMATION HIGHWAYS (1994).

0578   **Electronic Frontiers Foundation.** c/o On Technology, 155 2nd St., Cambridge, MA 02144. Voice: 617 864 0665. Fax: 617 864 0866. Internet: eff@eff.org. Public interest group, according to co-founder Mitchell Kapor (designer of Lotus 1-2-3), "dedicated to realizing the democratic potential of new computer and communications media." Promotes awareness of legal aspects of new information technologies.

0579   **Institute for Public Representation.** Georgetown University Law Center. 600 New Jersey Ave. NW, Washington, DC 20001. Voice: 202 662 9535. Fax: 202 662 9681. Public interest law firm specializing in communications regulatory policy; helps citizens' groups make input unto local and federal electronic media regulation. Publishes annual review of communications cases.

0580   **Internet Society** 1895 Preston White Dr., Ste. 100, Reston, VA 22091. Voice: 703 648 9888. Fax: 703 620 0913. Internet: isoc@isoc.org. "Non-profit organization" to promote technical development of the Internet to the benefit of academic, educational, scientific, industrial communities and the public at large. Helps to plan the Internet's growth. Publishes newsletter (1992-date).

0581   **Telecommunication Research and Action Center.** P.O. Box 12038. Washington, DC 20005. Voice: 202 462 2520. Fax: 202 408 1134. Nonprofit consumer interest group in telecommunications issues and clearinghouse for information on media reform. Publishes newsletter TELETIPS (1990-date, semiannual) as well as monographs.

4-E-4. Consultants Directories

0582   **CONSULTANTS AND CONSULTING ORGANIZATIONS DIRECTORY.** Detroit: Gale, 1966–date, annual. Online version available from HRIN (Human Resources Information Network): covers current edition; updated semiannually. Classified descriptions include personnel, addresses, telephone and fax numbers, with brief notes on consulting activities and publications. Chapter covering "Computer Technology, Telecommunications, and Information Services." Supplemented by NEW CONSULTANTS (1973–date).

0583   **DUN'S CONSULTANTS DIRECTORY.** Parsippany, NJ: Dun's Marketing Service, 1985–date, annual. CD-ROM version available in DUN'S BUSINESS LOCATOR from Dun and Bradstreet Information Services: covers current edition; updated semiannually. Directory of U.S. consulting firms. Geographic, consulting area/speciality, branch offices indexes.

See 0038. Krol, **TELECOMMUNICATIONS DIRECTORY.**

0584   **Society of Telecommunications Consultants.** 23123 S. State Rd. 7, Suite 220, Boca Raton FL 33428. Voice: 407 852 7071. Fax: 407 852 9262. Toll free: 800 782 7670. Telecommunications professionals and companies concerned with self-regulation of industry. Represents membership on legislative, regulatory, and commercial issues. Maintains Vendor Advisory Council which provides information to consultants. Publishes CONSULTANT MEMBERSHIP DIRECTORY (1984–date, annual) and newsletter.

4-F. Selected Secondary Resources

4-F-1. Industry Surveys

0585   Alleman, James H., and Richard D. Emmerson. **PERSPECTIVES ON THE TELEPHONE INDUSTRY: THE CHALLENGE FOR THE FUTURE.** New York: Harper & Row, 1989, 324 pp. Twenty-two chapters assess the industry's economics, regulatory risk to service providers, monopoly risk in the marketplace, price drivers, cost allocation drivers, and technology drivers. Figures, tables, index.

0586   Bradley, Stephen P., and Jerry A. Hausman, eds. **FUTURE COMPETITION IN TELECOMMUNICATIONS.** Boston: Harvard Business School Press, 1989, 340 pp. Ten chapters in sections on industry perspectives, competition in existing markets, special needs of large users, and electronic information services. Tables, notes, index.

0587   Huber, Peter. **THE GEODESIC NETWORK: 1987 REPORT ON COMPETITION IN THE TELEPHONE INDUSTRY.** Washington: GPO (Department of Justice), 1987, ca 300 pp. The most detailed post-divestiture overview of all aspects of the industry—done for the first (and only) triennial review of the MFJ. Tables, charts, diagrams, notes.

0588   _____, et al. **THE GEODESIC NETWORK II: 1993 REPORT ON COMPE-TITION IN THE TELEPHONE INDUSTRY.** Washington, DC: Geodesic Co, 1992, ca 400 pp. Tables and charts supplement chapters that update much of the 1987 volume. Color maps show RBOC ownership changes of cellular franchises. Notes, maps, tables.

0589   National Telecommunications and Information Administration. **THE NTIA INFRA-STRUCTURE REPORT: TELECOMMUNICATIONS IN THE AGE OF INFORMATION.** Washington: GPO, 1991. 316 pp. Detailed analysis of telecommunication industry structure and operations. Tables, charts.

0590   Weinhaus, Carol, and Anthony Oettinger. **BEHIND THE TELEPHONE DEBATES.** Norwood, NJ: Ablex, 1988, 253 pp. Diagrams and text on the economic and policy issues after divestiture making good use of original diagrams. Tables, notes, index.

4-F-2. Economic Analyses

0591   Allison, John R., and Dennis L. Thomas, eds. **TELECOMMUNICATIONS DEREGU-LATION: MARKET POWER AND COST ALLOCATION ISSUES.** Westport, CT: Quorum, 1990, 283 pp. Six economic papers and related discussions. Notes, charts, index.

0592   Antonelli, Christiano, ed. **THE ECONOMICS OF INFORMATION NETWORKS.** Amsterdam: North-Holland, 1992, 477 pp. Some 20 papers based on American and overseas experience with network design and operation, network economics, network strategies, and continuing network evolution. Notes, tables, index.

0593   Baughcum, Alan and Gerald R. Faulhaber, eds. **TELECOMMUNICATIONS ACCESS AND PUBLIC POLICY.** Norwood, NJ: Ablex, 1984, 24 pp. Papers from a 1982 conference concerned with local access questions. Some 18 chapters appear in sections on technology, markets, recent social policy, and the regulatory agenda. Index.

0594   Egan, Bruce L. **INFORMATION SUPERHIGHWAYS: THE ECONOMICS OF ADVANCED PUBLIC COMMUNICATION NETWORKS.** Norwood, MA: Artech, 1991, 188 pp. Economic, regulatory, and corporate aspects of broadband networking. Bibliography, index.

0595   Hyman, Leonard S., et al. **THE NEW TELECOMMUNICATIONS INDUSTRY: EVOLUTION AND ORGANIZATION.** Arlington, VA: Public Utility Reports, 1987 (two vols.), 473 + 204 pp. Technology, organization, regulation and financial basics of the industry. Tables, charts, selected bibliography.

0596   Wenders, John. **THE ECONOMICS OF TELECOMMUNICATIONS: THEORY AND POLICY.** Cambridge, MA: Ballinger, 1986, 296 pp. Efficiency, a framework for analyzing telecommunication markets, optimal pricing, theory of economic regulation, deregulating toll markets, and the like. Notes, charts, index.

4-F-3. Telecommunication Management

0597   Belitsos, Byran, and Jay Misra. **BUSINESS TELEMATICS: CORPORATE NETWORKS FOR THE INFORMATION AGE.** Homewood, IL: Dow Jones-Irwin, 1986, 460 pp. Integrates several aspects: technical basics, office communication networks, corporate and public networks, a series of application case studies, and policy concerns. Diagrams, notes, index.

0598   Briere, Daniel D. **LONG DISTANCE SERVICES: A BUYER'S GUIDE.** Norwood, MA: Artech, 1990, 293 pp. The market, services, carriers, and management decisions needed to integrate the whole. Diagrams, bibliography, index.

0599   Elbert, Bruce R. **NETWORKING STRATEGIES FOR INFORMATION TECHNOLOGY.** Norwood, MA: Artech, 1992, 257 pp. Designed for system managers, its coverage includes disaster recovery, network security management, and keeping up to date.

0600   Gasman, Lawrence. **MANAGER'S GUIDE TO THE NEW TELECOMMUNICATIONS NETWORK.** Norwood, MA: Artech, 1988, 239 pp. Good basic guide assuming little prior knowledge. Index.

0601   Keen, Peter G.W. **COMPETING IN TIME: USING TELECOMMUNICATIONS FOR COMPETITIVE ADVANTAGE.** Cambridge, MA: Ballinger, 1988 (2nd ed.), 302 pp. The most advantageous ways to apply telecommunications as an integral part of corporate strategy. Index.

0602   Muller, Nathan J. **MINIMUM RISK STRATEGY FOR ACQUIRING COMMUNICATIONS EQUIPMENT AND SERVICES.** Norwood, MA: Artech, 1989, 457 pp. Choosing and operating systems and services and pitfalls to avoid. Index.

0603   _____, and Robert P. Davidson. **LANs to WANs: NETWORK MANAGEMENT IN THE 1990s.** Norwood, MA: Artech, 1990, 540 pp. Detailed guide to all aspects of equipment and service ordering and management. Diagrams, acronyms, index.

0604   Office of Management and Budget. **CURRENT INFORMATION TECHNOLOGY RESOURCE REQUIREMENTS OF THE FEDERAL GOVERNMENT.** Washington: GPO, 1990–date, annual. 3rd annual report (FY 94) gives data for 1991–1998 federal agencies information technology acquisitions plans submitted in support of BUDGET OF THE U.S. GOVERNMENT. Indexed in STATISTICAL MASTERFILE (0513).

0605   _____. **INFORMATION RESOURCES MANAGEMENT PLAN OF THE FEDERAL GOVERNMENT.** Washington: GPO, 1984–date, annual. Annual report of federal agencies' information technology spending and information management activities: data on computer and telecommunications hardware, software, services, computer security, EDI, etc. Indexed in STATISTICAL MASTERFILE (0513).

0606   Pruitt, James B. **TELECOMMUNICATION PROJECT MANAGEMENT.** Chicago: Telephony, 1987, 130 pp. Step-by-step process, with several useful case studies.

0607   Schlesinger, Leonard A. et al. **CHRONICLES OF CORPORATE CHANGE: MANAGEMENT LESSONS FROM AT&T AND ITS OFFSPRING.** Lexington, MA: Lexington Books, 1987, 251 pp. Growing out of the divestiture, these 10 chapters assess management under stress. Figures, tables, glossary, notes, index.

0608   Sheth, Jagdish N. and Gary Frazier, eds. **ADVANCES IN TELECOMMUNICATIONS MANAGEMENT.** Greenwich, CT: JAI Press, 1990, 3 vols. Series of subject-specific volumes on different aspects of overall topic. First dealt with research and development marketing, the second with evolution of procurement in telecommunications.

1 MANAGING THE R&D/MARKETING INTERFACE FOR PRODUCT SUCCESS: THE TELECOMMUNICATIONS FOCUS, 216 pp.

2 PURCHASING IN THE 1990'S: THE EVOLUTION OF PROCUREMENT IN THE TELECOMMUNICATIONS INDUSTRY, 257 pp.

3 INFORMATION TECHNOLOGY AND CRISIS MANAGEMENT, 224 pp.

0609   Terplan, Kornel. COMMUNICATION NETWORKS MANAGEMENT. Englewood Cliffs, NJ: Prentice Hall, 1992 (2nd ed.), 702 pp. Critical success factors, networking trends, generic architecture of a network, information compression and processing, fault management, performance management, security, accounting, network capacity planning, etc. Tables, charts, bibliography, index.

Applications/Impact

Our fifth chapter includes resources that reflect telecommunications' wide application and many impacts in many areas of daily life, including privacy, banking, education, shopping, and work. In addition to bibliographies and directories, we have included selected corporate, academic, and private organizations that conduct research on telecommunications' economic, social, political, and cultural implications. Selected secondary resources include studies in telecommunications' impact on business, the individual, and American society and culture.

5-A. Bibliographic Resources

5-A-1. Bibliographies

0610   Abshire, Gary M. **IMPACT OF COMPUTERS ON SOCIETY AND ETHICS: A BIBLI-OGRAPHY.** Morristown, NJ: Creative Computing, 1980, 120 pp. Alphabetical listing of 1,920 items (books, scholarly and popular articles) from 1948–1979 on social implications of electronic technologies. Covers information privacy and security, computer crime, etc.

0611   Boudreaux, Sybil A., Marilyn L. Hankel, and Terrence E. Young, Jr. **ELECTRONIC BANKING: AN ANNOTATED BIBLIOGRAPHY, 1978–1983.** Chicago: Council of Planning Librarians, 1985, 68 pp. (CPL Bibliography, 155). About 500 classified, briefly annotated references of trade, business, and management journal articles on electronic funds transfer. Focuses on ATMs, industry applications, international, rules and regulations.

See 0911. **BRIDGING THE GAP III.**

0612   Bureau of Labor Statistics. **PRODUCTIVITY: A SELECTED, ANNOTATED BIBLIOG-RAPHY, 1983–87.** Washington: GPO, 1990, 160 pp. (BULLETIN 2360). Classified annotated listing of over 1,000 items (books, articles, dissertations), with particularly useful coverage of electronic and information technology's influence on the full rage of industries (including the computer and telecommunication industries). Author and subject indexes. Updates similar BLS bibliographies 1226 (1958), 1514 (1966), 1776 (1971), 1933 (1977), 2051 (1980), and 2212 (1984).

0613   Burge, Elizabeth J. **COMPUTER MEDIATED COMMUNICATION AND EDUCATION: A SELECTED BIBLIOGRAPHY.** Toronto: Distance Learning Office, Ontario Institute for Studies in Education, 1992, 87 leaves. Some 300 unannotated entries in separate alphabetical and classified listings. Identifies studies on distance learning's student/participant and teacher/moderator perspectives, messaging, tools and techniques, and educational and non-educational contexts. Emphasis on implications and applications in classroom, vocational and job training, health and medical education.

0614   Casey, Verna. **ELECTRONIC DETENTION-HOUSE ARREST AS A CORREC-TIONAL ALTERNATIVE: A SELECTED BIBLIOGRAPHY.** Monticello, IL: Vance Bibliographies, 1988, 7 pp. (Public Administration Series, P–2555). References on electronic monitoring.

0615   Clatanoff, Robert M. **AD VALOREM ASSESSMENT OF TELECOMMUNICATIONS PROPERTY: A BIBLIOGRAPHY, DIRECTORY, AND RESOURCE GUIDE.** Chicago: Council of Planning Librarians, 1982, 32 pp. (CPL Bibliography 83). Focuses on telcommunications as revenues source for local government. Reviews literature (some 150 references) of telecommunications devices as equipment to be assessed for property tax in relation to decisions on land use and land use regulation. Directory of telecommunications organizations and government agencies.

0616   Cutcliffe, Stephen H., Judith A. Mistichelli, and Christine M. Roysdon. **TECHNOLOGY AND VALUES IN AMERICAN CIVILIZATION**. Detroit: Gale, 1980, 704 pp. American Studies Information Guide Series, 9. Classified annotated listing of literature addressing the "role of technology in American culture" (p. xv), both American attitutes toward technology as well as technology's influence and effects on the formation of values. "Communications" (pp. 313–331) covers telegraph, telephone, radio, television, and computers. Author, title, subject indexes. Many relevant cross references.

0617   Dillman, Don A., and Jurg Gerber. **SOCIOLOGICAL IMPLICATIONS OF INFORMATION TECHNOLOGY: A BIBLIOGRAPHY OF RECENT PUBLICATIONS**. Chicago: Council of Planning Librarians, 1985, 32 pp. (CPL Bibliography 170). 530 references on cable television, computer crime, electronic mail, etc. Detailed subject index.

0618   Gold, John Robert, and Michael Baker. **TELECOMMUNICATIONS AND URBAN SPACE: A SELECTED BIBLIOGRAPHY**. Monticello, IL: Vance Bibliographies, 1979, 16 pp. (Public Administration Series, P–362). About 150 unannotated references in topical sections for telecommunications and "the future city"; intra-urban "spatial interdependencies"; applications (such as telecommuting); and telecommunications and government and public services.

0619   Grayson, Lesley. **THE SOCIAL AND ECONOMIC IMPACT OF NEW TECHNOLOGY, 1979–1984: A SELECT BIBLIOGRAPHY**. Letchworth, England: Technical Communications, 1984, 80 pp. Classified, annotated listing of about 700 books and articles on "social and economic impacts" of new electronic technologies on government, industry, business and home. Especially strong coverage of employment issues. Chapters cover "Homeworking," "Office automation," "Viewdata systems and cable television," and other topics. Unindexed.

0620   Henwood, Felicity, and Graham Thomas. **SCIENCE, TECHNOLOGY, AND INNOVATION: A RESEARCH BIBLIOGRAPHY**. Brighton, England: Harvester, 1984, 250 pp. Classified arrangement of about 2,000 unannotated entries on impact of technology on society, environment, government, and work. Useful keyword index cross references entries under "Communications," "Electronics," "Information technology," "Microelectronics," "Telecommunications."

0621   Hill, George H. **BLACK MEDIA IN AMERICA: A RESOURCE GUIDE**. Boston: G.K. Hall, 1984, 333 pp. See listings under "Cable television" for studies of minority cable station ownership. Author and subject indexes.

0622   Hudson, Heather E. **A BIBLIOGRAPHY OF TELECOMMUNICATIONS AND SOCIO-ECONOMIC DEVELOPMENT**. Norwood, MA: Artech, 1988, 241 pp. Some 1,124 entries with annotations consisting of descriptive tags (like "World," "Telephone," "Policy"). Cross referencing indexes for "Geographical Regions," "Technologies and Media," "Content and Applications," and "Organizations." An important bibliography.

0623   Lent, John. **WOMEN AND MASS COMMUNICATION: AN INTERNATIONAL ANNOTATED BIBLIOGRAPHY**. New York: Greenwood, 1991, 481 pp. Detailed subject index identifies relevant entries under "Computers," "Technology," "Telecommunications," "Telephone operators," and others.

0624   Maddux, Cleborne D. **DISTANCE EDUCATION: A SELECTED BIBLIOGRAPHY**. Englewood Cliffs, NJ: Educational Technology Publications, 1992, 71 leaves. Some 600 unannotated entries for books, articles, and ERIC documents in broadly classified sections for general studies, "Problems and Cautions" (budgets, programming), research, U.S. and international projects, and "Issues and Trends" (racial and ethnicity, gender, socioeconomic, cross-cultural considerations). Emphasis on electronic learning and interactive media.

0625   **MARXISM AND THE MASS MEDIA: TOWARDS A BASIC BIBLIOGRAPHY.** New York: International Mass Media Research Center, 1972–1980, 7 vols. in 3. "Global, multilingual, annotated bibliography of Marxist studies on all aspects of communications." 825 entries: vols. 1, 2, 3, entries 1–453, "revised edition" (1978); vols. 4, 5, entries 454–658 (1976); vols. 6, 7, entries 659–825 (1980). Relevant listings under "Cable TV," "Computers," "Electronic surveillance," "Satellites," "Telephone." Subject, author, and country indexes.

0626   Oman, Ray C. **ORGANIZATION PRODUCTIVITY AT THE CROSSROADS: A PARTIALLY ANNOTATED BIBLIOGRAPHY.** Monticello, IL: Vance Bibliographies, 1989, 54 pp. (Public Administration Series, P–2615). About 200 classified annotated references on use of automation, computers, and information technology in industry and government, with comparative international focuses (U.S., Japan, Germany, etc.).

0627   _____, and Nilda D. Godwin. **AN ASSESSMENT OF INFORMATION RESOURCE MANAGEMENT OVER A DECADE: AN ANNOTATED BIBLIOGRAPHY.** Monticello, IL: Vance Bibliographies, 1991, 99 pp. (Public Administration Series, P–3115). About 300 annotated references on influence of IT and IRM on organizational management, planning and policy, technology, innovation and change, etc.

0628   Peterson, Lorna. **ELECTRONIC FUNDS TRANSFER SYSTEMS.** Monticello, IL: Vance Bibliographies, 1988, 7 pp. (Public Administration Series, P–2404). About 100 unannotated references.

0629   Romiszowski, Alexander J. **COMPUTER MEDIATED COMMUNICATION: A SELECTED BIBLIOGRAPHY.** Englewood Cliffs, NJ: Educational Technology Publications, 1992, 55 pp. (Educational Technology Selected Bibliography Series, 5). About 500 classified unannotated entries for book, articles, dissertations, proceedings, and federal and state documents. Chapters cover general surveys and studies of trends and policies; "Techniques of Use"; "Tools" (hardware, software, systems, and logistics); "Networks and Networking"; distance applications in education, government, industry; and collaborative and cooperative work. Unindexed.

0630   Sanchez, James Joseph. **ELECTRONIC BULLETIN BOARDS AND COMPUTER CONFERENCING: A SELECTIVE, ANNOTATED BIBLIOGRAPHY.** Monticello, IL: Vance Bibliographies, 1987, 9 pp. (Public Administration Series, P–2258). About 30 annotated items on U.S. and international projects.

0631   _____. **ELECTRONIC MAIL: A SELECTIVE, ANNOTATED BIBLIOGRAPHY.** Monticello, IL: Vance Bibliographies, 1987, 10 pp. (Public Administration Series, P–2259). About 40 annotated items—trade journals, published proceedings, and ERIC documents.

0632   _____. **THE NEW TELECOMMUNICATIONS TECHNOLOGIES AND WOMEN IN THE WORK FORCE: A SELECTIVE, ANNOTATED BIBLIOGRAPHY.** Monticello, IL: Vance Bibliographies, 1987, 7 pp. (Public Administration Series, P–2264). About 25 annotated items on women and office automation and telecommuting.

0633   _____. **SATELLITE-BASED EDUCATIONAL INFRASTRUCTURE IN ALASKA: THE EDUCATIONAL SATELLITE COMMUNICATIONS DEMONSTRATION: A SELECTIVE, ANNOTATED BIBLIOGRAPHY.** Monticello, IL: Vance Bibliographies, 1987, 8 pp. (Public Administration Series, P–2267). About 30 annotated items from journals and ERIC documents on distance education project.

0634   _____. **TELECOMMUTING: A SELECTIVE, ANNOTATED BIBLIOGRAPHY.** Monticello, IL: Vance Bibliographies, 1987, 15 pp. (Public Administration Series, P–2263). About 60 annotated items from trade and management journals.

0635   _____. **TELECONFERENCING SERVICES AND SYSTEMS: A SELECTIVE, ANNOTATED BIBLIOGRAPHY.** Monticello, IL: Vance Bibliographies, 1987, 8 pp. (Public Administration Series, P–2260). About 25 annotated references to trade journal articles.

0636 _____. **TELESHOPPING, TELEBANKING, AND TELESOFTWARE: SERVICES AND SYSTEMS: A SELECTIVE, ANNOTATED BIBLIOGRAPHY.** Monticello, IL: Vance Bibliographies, 1987, 12 pp. (Public Administration Series, p–2261). About 50 annotated references to trade journal articles that describe specific pilot projects.

0637 _____. **VIDEOCONFERENCING SERVICES AND SYSTEMS: A SELECTIVE, ANNOTATED BIBLIOGRAPHY.** Monticello, IL: Vance Bibliographies, 1987, 6 pp. (Public Administration Series, P–2266). About 25 annotated items, mostly from trade journals.

0638 Schneider, Jerry B., and Anita M. Francis. **AN ASSESSMENT OF THE POTENTIAL OF TELECOMMUTING AS A WORK-TRIP REDUCTION STRATEGY: AN ANNOTATED BIBLIOGRAPHY.** Chicago: Council of Planning Librarians, 1989, 40 pp. (CPL Bibliography 246). Reviews some 100 references on employee, management, social, legal and regulatory, cost-benefit, and technological issues of telecommuting. Directory of professionals and organizations in telecommuting research.

0639 Shearer, Benjamin F., and Marilyn Huxford. **COMMUNICATION AND SOCIETY: A BIBLIOGRAPHY ON COMMUNICATIONS TECHNOLOGIES AND THEIR SOCIAL IMPACT.** Westport, CT: Greenwood, 1983, 242 pp. Arranges unannotated entries for 2,732 books and articles in chapters for development, influence, social impact, and politics, with subordinate listings, including "Telegraph and Cable," "Telephone," "Radio and Wireless Telegraphy." Useful author and subject indexes: cross references to "Telecommunications" and many "tele-" words, as well as "Cable television," "Satellites," etc.

0640 Shervis, Katherine. **SATELLITE TELECONFERENCING: AN ANNOTATED BIBLIOGRAPHY.** Madison: EDSAT Center, Space Science and Engineering Center, University of Wisconsin, 1972, 130 pp. Annotated entries for about 200 books, documents, and articles. Author, permuted subject index.

0641 Warf, Barney. **BIBLIOGRAPHY OF TELECOMMUNICATIONS AND REGIONAL DEVELOPMENT.** Chicago: Council of Planning Librarians, 1989, 10 pp. (CPL Bibliography 242). About 100 briefly annotated references, mostly to economics, geography, and public policy journal articles. Focuses include teleports, office location, and banking.

0642 White, Anthony G. **911 EMERGENCY COMMUNICATIONS SYSTEMS: A SELECTED BIBLIOGRAPHY.** Monticello, IL: Vance Bibliographies, 1986, 6 pp. (Public Administration Series, P–2065). About 50 unannotated references to popular journal articles and city and regional publications.

0643 Whitehouse, Martha. **MICROELECTRONIC TECHNOLOGY & AUTOMATION: EMPLOYMENT AND TRAINING ISSUES, A SELECTED BIBLIOGRAPHY.** Monticello, IL: Vance Bibliographies, 1988, 8 pp. (Public Administration Series, P–2331). Unannotated list of trade journal articles and government documents, 1964–1988, with employment projections through 2000, on communications and computer technologies' impact on labor.

0644 Zureik, Elia, and Dianne Hartling. **SOCIAL CONTEXT OF THE NEW INFORMATION AND COMMUNICATION TECHNOLOGIES: A BIBLIOGRAPHY.** New York: Peter Lang, 1987, 310 pp. (American University Studies: Series XV, Communications, 2). Alphabetical listing of over 6,500 items on "social impact of information and communication technology," referenced by over 50 subject codes (e.g., "PI" for "policy issues," "SA" for "satellite, space communications." Useful coverage of telemedicine, distance learning, telework and MIS, networking, and transborder data communications. Good coverage of U.S. and Canadian documents.

5-A-2. Selected Abstracting, Indexing, and Electronic Database Services

See 0492. **BUSINESS PERIODICALS INDEX.**

0645 **CURRENT INDEX TO JOURNALS IN EDUCATION** (CIJE). Phoenix, AZ: Oryx Press, 1969–date, monthly. Online version ERIC available from BRS, Dialog, Knowledge Index, OCLC:

covers 1966–date, updated monthly. CD-ROM version ERIC available from Silver Platter: covers 1966–date, updated quarterly. Sponsored by the Educational Resources Information Center (ERIC). Useful for coverage of scholarly and research literature on educational telecommunication and information technology. Abstracts articles, with author and subject indexes, that are published in about 750 journals, including AMERICAN JOURNAL OF DISTANCE EDUCTION and EC&TJ: EDUCATIONAL COMMUNICATION AND TECHNOLOGY JOURNAL. ERIC's complementary RESOURCES IN EDUCATION (RIE) (Washington: GPO, 1966–date) indexes other separately published ERIC documents. Electronic versions cumulate CIJE and RIE.

See 0020. **DISSERTATION ABSTRACTS INTERNATIONAL.**

0646 **INDEX MEDICUS.** Washington: National Library of Medicine, 1879–date, monthly. Many variant titles. Online versions MEDLINE (and variant titles) available from BRS, Dialog, LEXIS/ NEXIS (0728). Data-Star, Knowledge Index, OCLC, and others: coverage varies (1966–date on Dialog); updating varies. Numerous CD-ROM versions, including CD PLUS/MEDLINE available from CD Plus Technologies: covers 1966–date; updated monthly. Indexes the world's biomedical scientific, practical, and policy literature—some 3,500 journals. Offers valuable coverage of research on such topics as "Telemedicine" (a medical subject heading), medical information systems, "Emergency Medical Service Communications Systems," distance education in health sciences. Electronic versions cumulate INDEX MEDICUS, INDEX TO DENTAL LITERATURE, and INTERNATIONAL NURSING INDEX.

0647 **MANAGEMENT CONTENTS.** Philadelphia: Institute for Scientific Information, 1975– date, twice monthly. Online version available from Dialog, Data-Star: covers 1974–date, updated monthly. Similar to ISI's CURRENT CONTENTS services (0019), reprints tables of contents from over 300 professional and scholarly journals in management, finance, economics, accounting, marketing, business operations, organizational behavior, and public administration, with particularly strong coverage of important journals in electronics, telecommunication, and new communication technologies industries, like INFORMATION AGE (1102), and others that frequently focus on telecommunication, such as CALIFORNIA MANAGEMENT REVIEW and COLUMBIA JOURNAL OF WORLD BUSINESS. Useful indexing for names of corporations, such as British Telecom, AT&T, and IBM.

0648 **PSYCHOLOGICAL ABSTRACTS.** Arlington, VA: American Psychological Association, 1927–date, monthly. Online version PSYCHLIT available from BRS, Dialog, Knowledge Index, OCLC (PSYCFIRST): covers 1967–date, updated monthly. CD-ROM version PSYCHLIT available from Silver Platter: covers 1974–date, updated quarterly. The major index in psychology. Useful indexing for scholarly and research literature on human-machine interaction and influences of technology on culture and society. Indexes major journals in communication. Relevant subject headings for telecommunication and new communication technologies include "Automated Information Processing," "Computer Assisted Diagnosis," "Speech Processing (Mechanical)," "Synthetic Speech," and "Telecommunications Media."

0649 **SOCIAL SCIENCE CITATION INDEX.** Philadelphia: Institute for Scientific Information, 1970–date, 3/year. Online version SOCIAL SCISEARCH available from BRS, Dialog: covers 1972–date, updated twice monthly. CD-ROM version SOCIAL SCIENCES CITATION INDEX COMPACT DISC EDITION available from Institute for Scientific Information: covers 1986–date, updated quarterly. Interrelated source, citation, and "Permuterm" indexes analyse a wide range of scholarly and professional journals. Coverage particulary strong in marketing, management, and media uses and effects. Permuterm headings for "Telecommunication" include "Advantages," "Development," "Economics," "Impacts," and "Regulations," among others.

0650 **SOCIAL SCIENCES INDEX.** New York: H. W. Wilson, 1974–date, quarterly. Online version available from Wilsonline, OCLC (SOCSCIIND): covers February 1984–date, updated twice weekly. CD-ROM version available from H. W. Wilson and Silver Platter: covers 1984–date, updated quarterly. Author and subject indexing of major scholarly journals in psychology, sociology, economics, political science, and education. Industry coverage from perspectives of current events and business; addresses educational, psychological, political, and sociological applications of new communication technologies. Indexes core telecommunication research journals. Provides important coverage of journals that frequently publish commentary and research on telecommunication, such as ECONOMIST, AMERICAN ECONOMIC REVIEW, FAR EASTERN ECONOMIC REVIEW, and JOURNAL OF POLICY ANALYSIS AND MANAGEMENT. Relevant headings include "Automatic Speech Recognition," "Cellular Radio," "Direct Broadcasting Satellite Television," "Electronic Mail Systems," "Facsimile Transmission," "Imaging Systems," "Optical Storage Devices," "Telephone Companies—Advertising," and "Videotelephones."

0651   **SOCIOLOGICAL ABSTRACTS.** San Diego, CA: Sociological Abstracts, 1952–date, 5/year. Online version SOCIOFILE available from BRS, Dialog, Knowledge Index, OCLC (SOCIOABS): covers 1963–date, updated 5/year. CD-ROM version SOCIOFILE available from Silver Platter: covers 1974–date, updated 3/year. Useful for analyses of scholarly research literature on media effects and the role of media in society and culture. Indexes significant telecommunication research journals. Detailed subject indexing with cross references. Useful telecommunication and new communication technologies headings include include "Automation," "Computers," "Electronic Technology," "Information Technology," "Networks," "Technological Innovations," "Technological Progress," "Telecommunications," and "Telephone Communications." Source list. Supplementary IRPS (International Review of Publications in Sociology) service lists and abstracts books and book reviews, with author and source indexes.

5-A-3. Directories

See 0038. Krol, **TELECOMMUNICATIONS DIRECTORY.**

See 0732. **GOVERNMENT RESEARCH DIRECTORY.**

See 0039. **RESEARCH CENTERS DIRECTORY.**

5-B. Research Entities

5-B-1. Corporate Research Entities

0652   **Allied Signal, Inc.** 101 Columbia Rd. and Park Ave., Morristown, NJ 07960. Voice: 201 455 2000. Fax: 201 455 4807. Comprehensive basic and product-oriented research programs. Annual volume of research is $426 million. Telecommunications laboratories include:

    1 **Communications System Division.** 1300 E. Joppa Rd., Towson, MD 21240. Voice: 410 583 4305. Fax: 410 337 7485.

    2 **Mobile Communications Division.** 2920 Haskell, Lawrence, KS 66044. Voice: 913 842 0402.

0653   **American Telephone and Telegraph Co. (AT&T).** 550 Madison Ave., New York, NY 10022. Voice: 212 605 5500. Fax: 212 308 1820. Operates comprehensive proprietary and sponsored basic and product-oriented research and development programs in acoustics, telephony, speech, hearing, computer technology, communications, and other areas. Annual volume of research is $2.4 billion. Maintains laboratories nation- and world-wide, including:

    1 **AT&T Bell Laboratories.** 600 Mountain Ave., Murray Hill, NJ 07974. Voice: 908 582 3000. World-Wide Web: http://www.research.att.com. Maintains substantial publication program. Publishes AT&T TECHNICAL JOURNAL (1063) and AT&T TECHNOLOGY (1064). Also publishes BELL LABORATORIES TALKS AND PAPERS (1962-date), monographs like TELECOMMUNICATIONS TRANSMISSION ENGINEERING, 3rd ed. (1990), and training technical manuals and videos. Library publishes annual bibliography (1956-date). For published historical record of accomplishments over half a century, see 0253.

2 **Engineering Research Center.** Box 900, Princeton, NJ 08542-0900. Voice: 609 639 1234. Fax 609 639 2827.

0654 **Bell Communications Research, Inc. (BellCore).** 290 W. Mt. Pleasant Ave., Livingston, NJ 07039. Voice: 201 740 3000. Fax: 201 740 6887. World-Wide Web: http://bellcore.com. Provides telecommunications research and development services for the Regional Bell Operating Companies. Supports substantial monograph, serial, and audiovisual publication program. Publishes CATALOG OF TECHNICAL INFORMATION (1990-date, annual), which lists technical reports and other publications of RBOCs; BELLCORE EXCHANGE (1988-date, bimonthly); ANNUAL REVIEW (1986-date); and DIGEST OF TECHNICAL INFORMATION (1984-date, monthly). Research studies and reports include RESEARCH ON ADVANCED TELEVISION FOR BROADBAND ISDN (1990), DIGITAL WIRELESS ACCESS AND PERSONAL COMMUNICATIONS (1992), and NETWORK MANAGEMENT HANDBOOK (1990). Video publications include NEW TRANSPORT TECHNOLOGIES: WIRELESS COMMUNICATIONS (1994), NATIONAL ISDN (1991), and COMMUNICATING BY LIGHT: PRINCIPLES AND APPLICATIONS OF OPTICAL COMMUNICATIONS TECHNOLOGY (1990).

0655 **Communications Satellite Corporation (COMSAT).** 6560 Rock Spring Dr., Bethesda, MD 20817. Voice: 301 214 3682. Research and development of communications satellite systems. Laboratories include:

1 **COMSAT Laboratories.** 22300 Comsat Dr., Clarksburg, MD 20871. Voice: 301 428 4000. Fax: 301 428 4288.

0656 **General Electric Co.** 3135 Easton Turnpike, Fairfield. CT 06432. Voice: 203 373 2211. Comprehensive basic and product-oriented research includes electronics and communications. Annual volume of research is $1.4 billion. Laboratories include:

1 **GE Information Services.** 410 N. Washington St., Rockville, MD 20850. Voice: 301 340 4000. Fax: 301 340 4488.

0657 **GTE Corp.** 1 Stamford, CT 06901. Voice: 203 965 2000. Fax: 203 965 2277. Basic and product-oriented research on telecommunications and electronics. Laboratories include:

1 **GTE Laboratories, Inc.** 40 Sylvan Rd., Waltham, MA 02254. Voice: 617 890 8460. Fax: 617 980 9320.

0658 **International Business Machine Corp. (IBM).** Old Orchard Rd., Armonk, NY 10504. Voice: 212 265 8210. Fax: 914 765 4190. Comprehensive basic and product-oriented research programs on communications, computers, and information technologies. Annual volume of research is $4.9 billion. Major laboratories for telecommunications research include:

1 **Watson Research Center.** Box 218, Yorktown Heights, NY 10598. Voice: 914 945 2122. Fax: 914 945 2973.

2 **Almuden Research Center.** 650 Harry Rd., San Jose, CA 95120-6099. Voice: 408 927 3080. Fax: 408 927 2100.

0659 **ITT Corp.** 1330 Avenue of the Americas, New York, NY 10019. Voice: 212 248 1000. Fax: 212 688 5686. Comprehensive research and development programs for ITT products, including microwave, satellites, and telecommunications. Annual volume of research is $565 million. Laboratories include:

1 **ITT Federal Services Corp.** 1330 Inderness Dr., 1 Gateway Pl., Colorado Springs, CO 80910. Voice: 719 574 5850. Fax: 719 380 8309.

0660 **Litton Industries, Inc.** 360 N. Crescent Dr., Beverly Hills, CA 90210. Voice: 212 859 5000. Fax: 213 859 5940. Product-oriented research and development in electronics and telecommunications. Annual volume of research is $166 million.

0661 **Martin Marietta Corp.** 6801 Rockledge Dr., Bethesda, MD 20817. Voice: 301 897 6000. Fax: 301 897 6704. Comprehensive research and development programs cover electronics and communications. Annual volume of research is $212 million.

0662   **Motorola, Inc.** 1303 Algonquin Rd., Schaumburg, IL 60196. Voice: 708 397 5000. Fax: 708 576 8003. Research and development programs. Annual volume of research is $1 billion. Telecommunications laboratories include:

       1 **Communications Sector** (Schaumburg IL). Voice: 708 576 1000. Fax: 708 576 2702.

       2 **Cellular Phones Division.** 5555 N. Neach, Fr. Worth, TX 76137. Voice: 817 232 6000. Fax: 817 232 6148.

       3 **Communications Sector, Portable and Systems Group.** 8000 W. Sunrise Blvd., Plantation, FL 33322. Voice: 305 475 6308.

0663   **Rockwell International Corp.** 2230 E. Imperial Hwy., El Segundo, CA 90245. Voice: 213 647 5000. Fax: 213 647 5524. Annual volume of research is $487 million. Laboratories include:

       1 **Telecommunications Division.** 4311 Jamboree Rd., Box C, Newport Beach, CA 92658-8902. Voice: 714 833 4600.

0664   **3M Co.** 3M Ctr., St. Paul, MN 55101. Voice: 612 733 1100. Fax: 612 736 8261. Annual volume of research is $865 million. Major telecommunications research facilities include:

       1 **3M Information Imaging and Electronics Sector** (St. Paul, MN). Voice: 612 733 1288. Fax: 612 733 8774.

0665   **Unisys Corp.** Township Ln. and Union Meeting Rd., Blue Bell, PA 19422. Voice: 215 542 4011. Fax: 215 986 6850. Annual volume of research is $746 million.

5-B-2. University-Based Research Institutes

0666   Carnegie Mellon University. **Information Networking Institute.** Pittsburgh, PA 15213. Voice: 412 268 7195.

0667   Fordham University. **Donald McGannon Communications Research Center.** Bronx, NY 10458. Voice: 212 579 2693. Fax: 212 579 2708.

0668   Massachusetts Institute of Technology. **Media Lab Communications Research Program.** Cambridge, MA 02139. Voice: 617 253 6630. Fax: 617 258 6264. Founded and directed by Nicholas Negroponte. Annual budget of $10.5 million. Sponsored by more than 100 corporations world-wide.

0669   Michigan State University. **Communications Technology Laboratory.** East Lansing, MI 48824. Voice: 517 353 3794. Fax: 517 353 5498. Research on virtual reality, multimedia, and other information technologies. Results published in scholarly journals and monographs.

0670   University of Colorado. **Center for Advanced Research in Telecommunications.** Boulder, CO 80309. Voice: 303 492 8916. Fax: 303 492 1112. Research arm of University's academic Interdisciplinary Program in Telecommunications (at same voice and fax). Technology-oriented program with particular interest in rural telecommunications. Publishes irregular reports, including RURAL TELECOMMUNICATIONS PROJECTS AND RESOURCES (1994).

0671   University of Mississippi. **Center for Wireless Communications.** University, MS 38677. Voice: 601 232 7779. Fax: 601 232 7796. Formerly Center for Telecommunications. Offers graduate and undergraduate courses.

0672   University of Nebraska. **Center for Management of Information Technology.** Omaha, NE 68182. Voice: 402 554 2521. Fax: 402 554 3747. Until 1993 the International Center for Telecommunications Management, which published occasional studies and reports in "ICTM Research Paper Series" (1989-1993), including James Alleman's TELECOMMUNICATIONS PRICING PRIMER (1991), Lee Heeseok's SUPPORTING RURAL TELECOMMUNICATION NETWORKS VIA HUB CITIES (1992) and Michael J. O'Hara's PUBLIC MEETINGS VIA TELECOMMUNICATIONS (1992); and conference proceedings. New center has no publications to date.

0673   University of Pennsylvania. **Center for Communications and Information Sciences and Policy.** Philadelphia, PA 19104. Voice: 215 898 9494. Fax: 215 898 1130.

0674   University of Rhode Island. **Research Institute for Telecommunications Information Marketing (RITIM)**, College of Business Administration, Ballentine Hall, Kingston, RI 02882. Voice: 401 792 5065. Fax: 401 792 4312. Internet: RITIM@URIACC.URI.EDU. Conducts research on marketing, organizational, behavioral, and strategic aspects of telecommunications industries. Publishes symposia proceedings, "Working Papers" series, and "Research Reports," including CONSULTANTS AS GATEKEEPERS: A SURVEY OF CONSULTANTS' ROLES IN THE TELECOMMUNICATIONS PURCHASE DECISION PROCESS and THE RIGHT VOICE: USER RESEARCH FOR EFFECTIVE TELECOMMUNICATIONS MARKETING, among others.

0675   University of Texas. **Center for Research on Communications Technology and Society**. Austin, TX 78713-7389. Voice: 512 471 5826. Fax: 512 471 8500. Internet: comtech@bongo.cc.utexas.edu. Focuses on "social, economic, and policy impacts of the emerging information age," with particular interest in telemedicine and U.S.-Mexican telecommunications. Publishes research studies and reports, including Frederick Williams, Liching Sung, and Jerry Barnaby's INNOVATIVE TELECOMMUNICATIONS APPLICATIONS: EXAMPLES IN BUSINESS, PUBLIC INSTITUTIONS, AND INFORMATION SERVICES (1992); Susan G. Hadden's REGULATING CONTENT AS UNIVERSAL SERVICE (1991); and Williams and Mary Moore's TELEMEDICINE: A PLACE ON THE INFORMATION HIGHWAY (1994). Also publishes newsletter GATEWAY (1990-date, semiannual).

5-B-3. Other Entities

0676   **The Brookings Institution**. 1775 Massachusetts Ave. NW, Washington, DC 20036. Voice: 202 797 4029; 202 797 6236 (library). Fax: 202 797 6004. Supports vigorous research and publication program on wide range of economic, political, and social issues, including telecommunications and information technology. Published Robert W. Crandall's AFTER THE BREAKUP: U.S. TELECOMMUNICATIONS IN A MORE COMPETITIVE ERA (1991), Crandall and Kenneth Flamm's CHANGING THE RULES: TECHNOLOGICAL CHANGE, INTERNATIONAL COMPETITION, AND REGULATION IN COMMUNICATIONS (1989), NEW ETHICS IN THE COMPUTER AGE (1986), Flamm's THE TRANSFER OF ADVANCED TECHNOLOGY: RECENT TRENDS AND IMPLICATIONS FOR MEXICO (1986), and SEMICONDUCTORS AND THE WORLDWIDE SPREAD OF TECHNOLOGY (1971), among others.

0677   **Institute for Information Studies**. Northern Telecom, 200 Athens Way, Nashville, TN 37228-1803. Voice: 800 766 3995; 615 734 4525. Fax: 615 734 4109. Joint venture of telecommunications systems supplier Northern Telecom and non-profit Aspen Institute. Provides a forum for senior executives to engage in dialogue on critical strategic information issues. Publishes ANNUAL REVIEW (1990-date); also issued with volume titles, including UNIVERSAL TELEPHONE SERVICE: READY FOR THE 21st CENTURY (0699), A NATIONAL INFORMATION NETWORK: CHANGING OUR LIVES IN THE 21st CENTURY (0700), and THE KNOWLEDGE ECONOMY: THE NATURE OF INFORMATION IN THE 21st CENTURY (0701).

0678   **Institute for Telecommunication Sciences**. Bldg. 30, 325 S. Broadway, Boulder, CO 80303. Voice: 303 497 3500. Research arm of NTIA (0770.7). Publishes technical reports and monographs, distributed by NTIS (0770.6), including TELECOMMUNICATIONS RESEARCH AND ENGINEERING REPORT (1971–date), PUBLICATIONS ABSTRACTS (1979–date), R.J. Matheson's A PRELIMINARY LOOK AT SPECTRUM REQUIREMENTS FOR FIXED SERVICES (1993), and Charles Samora's AN EXTENDED DATABASE OF MICROWAVE COMMON CARRIER ANTENNA GAIN PATTERNS (1990).

5-C. Guides to Educational Programs

0679   Clemmensen, Jame M. **TELECOMMUNICATIONS EDUCATION: AN INFORMAL GUIDE AND ASSESSMENT**. Dallas: International Communications Association, 1986 (2nd ed.),

345 pp. Detailed information on 39 university-based programs, most of them graduate, and divided into interdisciplinary, integrated, and research programs.

0680   **EDUCATIONAL RANKINGS ANNUAL**. Detroit: Gale, 1994, 795 pp. Excerpts and cross references sources (books, articles, etc.) of ranking data in topical arrangement. Very convenient for identifying most well-recognized undergraduate, graduate, and professional programs, most frequently cited communications journals, scholars, etc.

0681   Elmore, Garland C., ed. **THE COMMUNICATIONS DISCIPLINES IN HIGHER EDUCATION: A GUIDE TO ACADEMIC PROGRAMS IN THE UNITED STATES AND CANADA**. Annandale, VA: Association for Communication Administration, 1993 (2nd ed.), 489 pp. Despite focus on speech and media courses, the increasing number of telecommunications courses and programs is reflected here.

0682   Not used.

0683   National Telecommunications Education Committee, **DIRECTORY OF TELECOMMU-NICATIONS SCHOOLS AND INSTITUTIONS**. Washington: North American Telecommunications Association, 1988 (2nd ed.), 118 pp. Lists professional and education institutions by province and state, associations and government organizations dealing with telecommunication, and offers an index by type of degree or certificate, and by name of institution.

0684   Sapolsky, Barry, ed. **DIRECTORY OF MEDIA PROGRAMS IN NORTH AMERICAN UNIVERSITIES AND COLLEGES, 1994-95**. Washington: Broadcast Education Assn., 1994, 127 pp. Includes many telecommunication courses among its detailed listings of faculty and facilities in nearly 200 programs.

0685   Sterling, Christopher H. **ICA GUIDE TO UNIVERSITY TELECOMMUNICATION PROGRAMS**. Dallas: International Communications Association, 1985, 32 pp. Brief annotated directory to 27 university academic programs (most in engineering and most on the masters level) plus 13 university research centers.

5-D. Selected Secondary Resources

5-D-1. Business Applications

0686   Cross, Thomas B., and Marjorie Raizman. **TELECOMMUTING: THE FUTURE TECHNOLOGY OF WORK**. Homewood, IL: Dow Jones-Irwin, 1986, 255 pp. Implementation of telecommuting progams, understanding human factors involved, etc. Glossary, bibliography, index.

0687   Fulk, Janet, and Charles Steinfield, eds. **ORGANIZATIONS AND COMMUNICATION TECHNOLOGY**. Newbury Park, CA: Sage, 1990. How changing technologies are impacting business and other organizations.

0688   Gray, Mike, et al. **TELEWORKING EXPLAINED**. Chichester, England and New York: Wiley, 1993, 289 pp. A British Telecom-sponsored study of the impact of telecommunications applications in the workplace: organizations and employers, individual teleworkers and home-based offices, economic factors, security and confidentiality concerns, and the technology itself. Appendices, bibliography, index.

0689   Hepworth, Mark E. **GEOGRAPHY OF THE INFORMATION ECONOMY**. New York: Guilford Press, 1990, 258 pp. Discusses uneven distribution of information and the impact of computer network innovations which may change that. Tables, charts, references, index.

0690   Kinsman, Francis. **THE TELECOMMUTERS**. Chichester, England, and New York: Wiley, 1987, 238 pp. Primarily based on British experience, this details the growing number of people who work at home and "telecommute." Chapters assess the impact of this process on workers, their families, managers, and work patterns generally. Appendices, index.

0691   Valovic, Thomas. **CORPORATE NETWORKS: THE STRATEGIC USE OF TELECOMMUNICATIONS**. Norwood, NJ: Artech, 1992, 156 pp. Chapters on technology and structure, how telecommunication is transforming the corporation, optimising internal corporate communications, public vs. private networks, case studies in strategic networking, etc. Diagrams, index.

0692   Zuboff, Shoshana. **IN THE AGE OF THE SMART MACHINE: THE FUTURE OF WORK AND POWER**. New York: Basic Books, 1988, 468 pp. Insightful review of past research and the author's own observation studies. Notes, index.

5-D-2. *Residential/Individual Applications*

0693   Mitchell, Bridger M. and Tenzing Donyo. **UTILIZATION OF THE U.S. TELEPHONE NETWORK**. Santa Monica, CA: Rand Corporation, 1994, 52 pp. Based in part on FCC data, some of which is reproduced here, this analyses changing patterns of consumer and business use. Tables, charts, notes, references.

0694   Rakow, Lana F. **GENDER ON THE LINE: WOMEN, THE TELEPHONE, AND COMMUNITY LIFE**. Urbana: University of Illinois Press, 1992, 165 pp. Detailed study of communication patterns in a mid-Western community. Bibliography, index.

0695   Wilson, Kevin G. **TECHNOLOGIES OF CONTROL: THE NEW INTERACTIVE MEDIA FOR THE HOME**. Madison: University of Wisconsin Press, 1988, 180 pp. Argues that the developing networks may actually lead to more government and business centralized control. References, index.

5-D-3. *Larger Social Concerns*

0696   Dilts, Marian May. **THE TELEPHONE IN A CHANGING WORLD**. New York: Longmans, 1941, 219 pp. Perhaps the first book-length assessment of telephone impact, this is an informal rather than research-based discussion. Notes, index.

0697   Dordick, Herbert S., and Georgette Wang. **THE INFORMATION SOCIETY: A RETROSPECTIVE VIEW**. Newbury Park, CA: Sage, 1993, 168 pp. Compares and contrasts experience in 19 different countries. Index.

0698   Guile, Bruce R., ed. **INFORMATION TECHNOLOGIES AND SOCIAL TRANSFORMATION**. Washington: National Academy Press, 1985, 173 pp. Six wide-ranging papers from the National Academy of Engineering.

0699   Institute for Information Studies. **UNIVERSAL TELEPHONE SERVICE: READY FOR THE 21st CENTURY?** Queenstown, MD: Aspen Institute, 1991, 184 pp. Six papers detailing changing definitions and applications of universal telephone service.

0700   _____. **A NATIONAL INFORMATION NETWORK: CHANGING OUR LIVES IN THE 21st CENTURY**. Queenstown, MD: Aspen Institute, 1992, 182 pp. Six papers on telecommunications infrastructure development.

0701   _____. **THE KNOWLEDGE INDUSTRY: THE NATURE OF INFORMATION IN THE 21st CENTURY**. Queenstown, MD: Aspen Institute, 1993, 184 pp. Six papers range widely on how concepts of information are changing.

0702   Office of Technology Assessment. **RURAL AMERICA AT THE CROSSROADS: NETWORKING FOR THE FUTURE**. Washington: GPO, 1991, 190 pp. How communications can aid the developmental process with case studies from four states. Tables, photos, diagrams, index.

0703   _____. **CRITICAL CONNECTIONS: COMMUNICATION FOR THE FUTURE**. Washington: GPO, 1990, 395 pp. Excellent assessment of the broader impact of society and policy implications. Tables, diagrams, notes.

0704   _____.**ELECTRONIC ENTERPRISES: LOOKING TO THE FUTURE**. Washington: GPO, 1994, 176 pp. Chapters on issues in electronic commerce, regulation, cooperative networking, promoting technology/industry developments, educating for technology transfer, and the roles of government and markets.

0705   Pool, Ithiel de Sola, ed. **THE SOCIAL IMPACT OF THE TELEPHONE**. Cambridge, MA: MIT Press, 1977, 502pp. Development of literature on, and the many social roles of the telephone. Notes, index.

0706   _____. **TECHNOLOGIES OF FREEDOM**. Cambridge, MA: Harvard University Press, 1983, 299 pp. In his last book, the long-time communications expert discusses some of the dangers, as well as some of the benefits of improving delivery systems. Notes, index.

0707   _____; edited by Eli M. Noam. **TECHNOLOGIES WITHOUT BOUNDARIES: ON TELECOMMUNICATIONS IN A GLOBAL AGE**. Cambridge, MA: Harvard University Press, 1990, 283 pp. Continuation of the book above, issued after Pool's death, and continuing his basically positive outlook. Notes, index.

0708   Sapolsky, Harvey M., Rhonda J. Crane, W. Russell Neuman, and Eli M. Noam. **THE TELECOMMUNICATIONS REVOLUTION: PAST, PRESENT, AND FUTURE**. New York: Routledge, Chapman & Hall, 1992, 217 pp. Collection of 19 academic papers in honor of the late Ithiel de Sola Pool, and taking a wide view of most aspects of telecommunications.

0709   Schement, Jorge Reina, and Leah Lievrouw, eds. **COMPETING VISIONS, COMPLEX REALITIES: SOCIAL ASPECTS OF THE INFORMATION SOCIETY**. Norwood, NJ: Ablex, 1988. Ten papers by different authors on varied aspects of the cover topic.

0710   Short, John, et al. **THE SOCIAL PSYCHOLOGY OF TELECOMMUNI- CATIONS**. New York: John Wiley, 1976, 195 pp. A pioneering effort, this summarizes then-available research on all aspects. References, indexes.

0711   Williams, Frederick. **THE NEW TELECOMMUNICATIONS: INFRASTRUCTURE FOR THE INFORMATION AGE**. New York: Free Press, 1991, 246 pp. A broad survey designed as a text, this relates technology to its various applications—and impacts. Glossary, references, index.

0712   _____, and John V. Pavlik, eds. **THE PEOPLE'S RIGHT TO KNOW: MEDIA DEMOCRACY, AND THE INFORMATION HIGHWAY**. Hillsdale, NJ: Lawrence Erlbaum, 1994, 258 pp. How changing information technologies can aid the democratic process, with some case study examples.

6

<hr>

Policy and Regulation

Chapter 6 details resources that describe the policy-making process in this field and organizations that make and implement U.S. telecommunications policy. This chapter covers the congressional, executive, independent and judicial portions of the federal government including the Office of Technology Assessment (0761); the National Telecommunications and Information Administration (0770.7) and Federal Communications Commission (0782), including the dramatic organizational changes of 1994. We have included descriptions of each of these federal legislative, regulatory, and judicial agencies plus selected state and multistate offices (as well as guides to them), in that the best information on the current status of a bill, regulation, or decision can be obtained by a phone call or faxed request. Selected secondary resources include surveys of policy, treatises and texts, and studies of access, privacy, and intellectual property, convergence of media and telecommunications, and—more recently—the National Information Infrastructure (NII).

6-A. Bibliographic Resources

6-A-1. Bibliographies

0713   Bennett, James R. **CONTROL OF THE MEDIA IN THE UNITED STATES: AN ANNOTATED BIBLIOGRAPHY**. New York: Garland, 1992, 819 pp. 4,749 briefly annotated entries in subject arrangement. Emphasis on broadcasting though including items on AT&T, cable, satellites, telecommunication regulation. Author and subject indexes.

0714   Casper, Dale E. **PUBLIC REGULATION OF INFORMATION SYSTEMS: RECENT LAWS AND REGULATIONS, JOURNAL ARTICLES, 1982–1989**. Monticello, IL: Vance Bibliographies, 1990, 5 pp. (Public Administration Series, P–2833). About 50 unannotated references to legal journal articles on government efforts to control information systems.

0715   Chin, Felix. **REGULATORY REFORM OF TELECOMMUNICATIONS: A SELECTED BIBLIOGRAPHY**. Monticello, IL: Vance Bibliographies, 1980, 50 pp. (Public Administration Series, P–521). About 300 annotated references to books, legal and economic journal articles, reports, and government documents.

0716   Flaherty, David H., ed. **PRIVACY AND DATA PROTECTION: AN INTERNATIONAL BIBLIOGRAPHY**. White Plains, NY: Knowledge Industry Publications, 1984, 286 pp. More than 1,800 items are listed—some are annotated—by topic with an author index.

0717   Goehlert, Robert U., and Fenton S. Martin. **POLICY ANALYSIS AND MANAGEMENT: A BIBLIOGRAPHY**. Santa Barabara, CA: ABC-Clio, 1985, 398 pp. Classified unannotated listing of policy analysis literature; relevant items in section for "Communication" (pp. 253–258) in chapter "Technological and Scientific Policy." Subject index cross references items under "Communication," "Computers," "Electronics," "Information," "Telecommunications," and others.

0718   Hunt, Marguerite J. **COPYRIGHT AND AUDIO/VISUAL MATERIALS: A BIBLIOGRAPHY OF LEGAL RAMIFICATIONS**. Monticello, IL: Vance Bibliographies, 1989, 7 pp. (Public

Administration Series, P–2576). About 80 unannotated references on software piracy and electronic property rights.

0719   Jung, Donald J. **IMPLICATIONS OF DEREGULATION ON THE PUBLIC ADMINIS-TRATION OF THE FEDERAL COMMUNICATIONS COMMISSION, 1981–1987: AN ANNOTATED BIBLIOGRAPHY.** Monticello, IL: Vance Bibliographies, 1989, 23 pp. (Public Administration Series, P–2629). 45 annotated references on how industry deregulation has affected structural changes in the FCC.

0720   Murin, William J., Gerald Michael Greenfield, and John D. Buenker. **PUBLIC POLICY: A GUIDE TO INFORMATION SOURCES**. Detroit: Gale, 1981, 283 pp. American Government and History Information Guide Series, 13. More than 1,248 classified annotated items; subject index identifies relevant items under "Federal Communications Commission," "Communications industry," "Communications," "Electronic surveillance," "Technology," and more.

0721   Nagel, Stuart S., editor. **BASIC LITERATURE IN POLICY STUDIES: A COMPRE-HENSIVE BIBLIOGRAPHY**. Greenwich, CT: JAI, 1984, 453 pp. Includes brief overviews and entensive classified unannotated bibliographies for "Communications Policy," by Jarol B. Manheim and Allison Ondrasik (pp. 233–47) and "Science Policy," by Alan L. Porter and Choon Y. Park (pp. 375–84).

0722   **PRIVACY AND SECURITY: A BIBLIOGRAPHY** and **1985 SUPPLEMENT** and **1986 SUPPLEMENT**. Washington: Computer and Business Equipment Manufacturers Association, 1980, 1985, 1986, 70 + 22 + 24 pp. Not annotated, but divided by topic.

0723   U.S. Superintendant of Documents. **TELECOMMUNICATIONS**. Washington: GPO, 1979–date, annual. Catalog listing of GPO publications for sale.

0724   Vance, Mary. **COMPUTER CRIME: REVISED EDITION OF P–1255.** Monticello, IL: Vance Bibliographies, 1988, 35 pp. (Public Administration Series, P–2355). About 400 unannotated references to trade journal articles and government documents.

0725   Zaffarano, Mark, and Anna Gnadt. **COURT TECHNOLOGY: A SELECTED BIBLIOG-RAPHY.** Monticello, IL: Vance Bibliographies, 1990, 7 pp. (Public Administration Series, P–2862). About 50 unannotated items on computers, fax machines, and other information systems in "the electronic courtroom."

6-A-2. Selected Abstracting, Indexing, and Electronic Database Services

0726   **INDEX TO LEGAL PERIODICALS.** New York: H. W. Wilson, 1908–date, monthly. Online version available from Wilsonline: covers August 1981–date, updated twice weekly. CD-ROM version available from H. W. Wilson: covers 1981–date, updated quarterly. Important for compre-hensive access to telecommunication's policy and legal English-language journal literature. Author and subject indexing of a broad range of legal periodicals published in the United States, Canada, Great Britain, Ireland, Australia, and New Zealand, as well as legal yearbooks and annual institutes and reviews, including CARDOZO ARTS & ENTERTAINMENT LAW JOURNAL (1108), HASTINGS COMMUNICATIONS AND ENTERTAINMENT LAW JOURNAL (1115), COMMUNICATIONS AND THE LAW (1109), COMPUTER/LAW JOURNAL (1111), FEDERAL COMMUNICATIONS LAW JOURNAL (1113), HIGH TECHNOLOGY LAW JOURNAL (1116), JOURNAL OF SPACE LAW (1117), LOYOLA ENTERTAINMENT LAW JOURNAL, and RUTGERS COMPUTER AND TECHNOLOGY LAW JOURNAL. Relevant strictly controlled subject headings include "Cable television," "Carriers," "Communications," "Computer crime," "Radio and television," "Telecommu-nications," "Telephones and telegraphs," and "Videotapes." Also includes headings for names of individuals and organizations. Appendixed "Table of Cases" indexes articles by litigants; "Table of Statutes" references popular and official names of laws.

0727   **LEGAL TRAC.** [CD–ROM.] Foster City, CA: Information Access, 1980–date, monthly. Online version LEGAL RESOURCE INDEX available from Dialog, LEXIS/NEXIS (0728),

WESTLAW (0729), Knowledge Index: covers 1980–date, updated monthly. References to law reviews, case notes, biographies, professional news, legislative analyses, bibliographies, and reviews in over 800 legal publications.

0728   **LEXIS/NEXIS.** [Online Service.] Dayton, OH: Reed Elsevier, 1973–date: coverage varies (newspaper, magazine coverage current five years); updating varies, dependent on publication, often same day or next day. Also available via Internet. Offers unmatched coverage of the full-texts of hundreds of primary and secondary materials in U.S., foreign national, and international law, business, medicine, and popular media. Full-texts arranged in major divisions and topical "libraries": all full-texts are searchable separately, in combination, and within libraries; many (not all) full-texts are available is several different divisions and libraries. LEXIS and related services cover legal, legislative, and regulatory information. Foremost of LEXIS's libraries for telecommunication research is FEDCOM (Federal Communications Library), including FCC decisions, reports, FCC DAILY DIGEST, FCC RECORD (well in advance of printed issues), federal case law related to communications, Title 47 of the U.S. CODE and CODE OF FEDERAL REGULATIONS, and selected industry and policy publications such as COMMUNICATIONS DAILY (1037) and FEDERAL COMMUNICATIONS LAW JOURNAL (1113). Also LEXIS's UTILITY (Public Utilities Law Library) contains state and federal statutes and case law on public utilities, commission decisions, and selected policy publications such as PUBLIC UTILTIES FORTNIGHTLY (1119). Other useful LEXIS libraries are LEGIS (federal and state legislative documents and information, including CONGRESSIONAL RECORD) and GENFED (CODE OF FEDERAL REGULATIONS, among over 100 files). NEXIS and related services cover news, business, and trade information. Important NEXIS libraries for telecommunication research are CMPCOM (Computers and Communications Library), covering trade and industry newsletters; ENTERT (Entertainment Library), containing industry newsletters, company profiles, and financial reports; and NEXIS (Nexis Library), including more than 650 domestic and foreign newspapers, newsletters, magazines, journals, wire services, and broadcast transcript services. Additional separate divisions useful for telecommunication industry information are COMPNY (Company Library), including investment bank market studies, SEC filings, and annual reports; and MARKET (Marketing Library), covering market research, consumer demographics, and product announcements and reviews. Also separate international libraries.

0729   **WESTLAW.** [Online Service.] Eagan, MN: West Publishing, 1978–date: coverage and updating varies, depending on file. Similar to LEXIS/NEXIS (0728), provides comprehensive bibliographic and full-text access to U.S. statutory, regulatory, and case law. The system includes WESTLAW COMMUNICATIONS LIBRARY file that specifically covers communications law, providing full-texts of federal and state legislative, administrative, and judicial documents, as well as full-texts of articles in law journals. Access to WESTLAW is typically limited to law libraries. Lacks strong coverage of popular media, trade newsletters featured in LEXIS/NEXIS.

6-B. U.S. Government

6-B-1. Bibliographic Resources

0730   **ENCYCLOPEDIA OF GOVERNMENTAL ADVISORY ORGANIZATIONS.** Detroit: Gale, 1973–date, irregular. 9th ed. (1994) for 1994/1995. Classified descriptions of 6,500 advisory groups. Listings under "tele"-words in alphabetical and keyword index.

0731   Low, Kathleen. **ELECTRONIC ACCESS TO GOVERNMENT AND GOVERNMENT INFORMATION: A SELECTIVE BIBLIOGRAPHY.** Monticello, IL: Vance Bibliographies, 1991, 7 pp. (Public Administration Series, P–3095). About 60 unannotated references on the "electronic city hall"—televising federal and state legislatures and community information services.

0732   **GOVERNMENT RESEARCH DIRECTORY.** Detroit: Gale, 1980–date, biennial. Formerly GOVERNMENT RESEARCH CENTERS DIRECTORY (1980–1982). 5th ed. for 1993/

1994 (1992). Online version RESEARCH CENTERS AND SERVICES DIRECTORY available from Dialog (see 0039). Arranged by federal branches, departments, and agencies. Relevant entries under departments of Commerce (NIST, NTIS, NTIA), Defense (Defense Information Systems Agency), NASA, etc. Geographic, subject, and master index. Useful listings under "Telecommunications," "Communications Systems," "Electronic Systems," and others.

0733   **GUIDE TO U.S. GOVERNMENT PUBLICATIONS.** Donna Andriot, editor. Mclean, VA: Documents Index, 1994, 1,637 pp. Earlier editions GUIDE TO U.S. GOVERNMENT SERIALS PUBLICATIONS. Commonly refered to as "Andriot," after John Andriot, original compiler. Gives complete bibliographic histories for some 35,000 U.S. government publications. Most useful for identifying early, pre-FCC serial publications on telegraphy, telephony, and other relevant topics. Arranged by government offices under SUDOC classes: see listings for FCC in "CC" class. Indexes for "Agency Class Chronology," agencies, titles, and keywords in titles: useful keywords include "Space," "Telecommunications," "Telegraph," "Telephone," and others.

0734   **MONTHLY CATALOG OF UNITED STATES GOVERNMENT PUBLICATIONS.** Washington GPO, 1895–date, monthly. Online version GPO MONTHLY CATALOG available from BRS, Dialog, OCLC, Tech Data, Wilsonline: covers July 1976–date, updated monthly. CD-ROM versions (titles vary) available from Autographics, H. W. Wilson, Information Access, Silver Platter: covers 1976–date, updating varies (monthly on Information Access). Most comprehensive bibliography of literature produced by the United States government, though often lagging months behind actual date of publication. Arranges document entries by issuing department, office, or agency (for example, the Department of Education, Corporation for Public Broadcasting, House Subcommittee on Telecommunications and Finance, NTIS, or FCC). Semiannual indexes of authors, titles, subjects, series/reports, contract numbers, stock numbers, and title-keywords (most useful for cross referencing "high" with "definition television" to identify hearings, reports, etc., otherwise listed by issuing office or committee).

0735   **OFFICIAL CONGRESSIONAL DIRECTORY.** Washington: GPO, 1809–date, annual. Originally intended as Congress's official guide to U.S. government offices. Detailed contents; name index.

0736   **U.S. GOVERNMENT MANUAL.** Washington: GPO, 1935–date, annual. Variant titles. Official guide to U.S. government offices. Name, agency/subject indexes.

0737   **WASHINGTON INFORMATION DIRECTORY.** Washington: Congressional Quarterly, 1975–date, annual. Basic data for U.S. government offices as well as Washington trade, industry, press, media, and other interest groups and organizations. Chapters covering "Communications and the Media" and "Space" include relevant listings.

6-B-2. Resources on Congress and Its Agencies

0738   Brightbill, George D. **COMMUNICATIONS AND THE UNITED STATES CONGRESS: A SELECTIVELY ANNOTATED BIBLIOGRAPHY OF COMMITTEE HEARINGS, 1870–1976.** Washington: Broadcast Education Association, 1978, 178 pp. More than 1,100 published hearings are listed, many dealing with telegraph, telephone, and later telecommunications concerns. Only guide of its kind.

0739   **CIS INDEX TO PUBLICATIONS OF THE UNITED STATES CONGRESS.** Washington: Adler, 1970–date, monthly. Online version available on Dialog: covers 1970–date, updated monthly. CD-ROM version CONGRESSIONAL MASTERFILE 2 available from Congressional Information Service: covers 1970–date, updated quarterly. Access to either print or electronic versions essential for research on all aspects U.S. domestic and international telecommunication. Indexes Congressional committee hearings and prints; House and Senate reports, documents, and special publications; Senate executive reports; and Senate treaty documents. Separate complementary abstract and index issues published monthly. "Index of Subjects and Names" (such as cable television, House Subcommittee on Telecommunication and Finance, and TeleCable Corporation) cross references uniquely numbered document abstracts, arranged under House and Senate committees. "Telecommunications" index heading "see also" references "Communications satellites," "Telephone and telephone industry," and alternatives. Other indexes for document titles; bill numbers; report numbers; document numbers; Senate hearing numbers; Senate print numbers; Superintendant of Documents numbers; and the names of Committee and Subcommittee chairs. Convenient annual third volume for "Legislative Histories" cumulates entries for hearings, debates, reports, etc., related to laws made within a particular year. Additionally, CIS also publishes CD-ROM CONGRESSIONAL MASTERFILE with retrospective files for the period 1789–1969, corresponding to CIS UNPUBLISHED U.S. SENATE COMMITTEE HEARINGS INDEX, 1823–1964 (1986), CIS U.S. CONGRESSIONAL COMMITTEE HEARINGS INDEX, 1833–1969 (1981), CIS U.S. CONGRESSIONAL COMMITTEE PRINTS INDEX, 1833–1969 (1980), and CIS U.S. SERIAL SET INDEX, 1789–1969 (1975): all are useful for historical research, particularly related to early telecommunication patents.

0740   **CONGRESS IN PRINT**. Washington: Congressional Quarterly, 1977–date, 48/year. Lists but does not annotate publications of CBO, OTA, and GAO as well as House and Senate committee hearings, reports, and committee prints.

0741   **CONGRESSIONAL QUARTERLY'S GUIDE TO CONGRESS**. Washington: Congressional Quarterly, 1991 (4th ed.), 836 + paged appendices. Standard guide to history, organization, and operations of Congress. Subject index identifies relevant sections on telecommunications.

0742   **CONGRESSIONAL STAFF DIRECTORY**. Mount Vernon, VA: Staff Directories, 1959–date, annual. 41st ed. for 1994. Identifies members of Congress. Useful for committee assignments. Keyword subject and name indexes. CD-ROM version STAFF DIRECTORIES ON CD-ROM available from Staff Directories; cumulates CONGRESSIONAL STAFF DIRECTORY, JUDICIAL STAFF DIRECTORY (0784), and FEDERAL STAFF DIRECTORY (0764): covers current editions, updated semiannually.

0743   Goehlert, Robert, and John R. Sayre. **THE UNITED STATES CONGRESS: A BIBLIOGRAPHY**. New York: Free Press, 1982, 376 pp. Classified unannotated listing of 5,620 items on the "history, development, and legislation process of Congress" (Introduction). Author and subject indexes; relevant listings under specific legislation, like "Communications Satellite Act of 1962," as well as under "Federal Communications Commission," "Science policy," "Telecommunications," and more.

6-B-3. Congress

Unless otherwise noted, addresses for Congress are RHOB: Rayburn House Office Bldg., Washington, DC 20515; and SHOB: Senate Hart Office Bldg., Washington, DC 20510. Internet access to House and Senate addresses: gopher://gopher.house.gov; and gopher://ftp.senate.gov.

6-B-3-i. Selected Congressional Committees

0744   **House. Committee on Appropriations. Subcommittee on Commerce, Justice, State, and Judiciary.** The Capitol, Washington, DC 20515. Voice: 202 225 3351. Legislation to appropriate funds for FCC, Board of International Broadcasting, and U.S. Information Agency.

0745   **House. Committee on Energy and Commerce. Subcommittee on Telecommunications and Finance.** 316 Ford House Office Bldg., Washington, DC 20515. Voice: 202 226 2424. Primary House center for legislation related to interstate and foreign communications (broadcast, cable, radio, wire, microwave, satellite, etc.) plus FCC oversight.

0746   **House. Committee on the Judiciary. Subcommittee on Intellectual Property and Judicial Administration.** Cannon House Office Bldg., Washington, DC 20515. Voice: 202 225 3926. Legislation on copyrights, patents, and trademarks.

0747   House. Committee on the Judiciary. Subcommittee on Economic and Commercial Law. RHOB. Voice: 202 225 2825. Legislation related to anticompetitive practices and monopolies in communications, including cable and telephone industry practices.

0748   House. Committee on Government Operations. Subcommittee on Government Information, Justice, Transportation, and Agriculture. RHOB. Voice: 202 225 3741. Jurisdiction over Freedom of Information Act. Oversees operation of Board for International Broadcasting.

0749   Senate. Committee on Appropriations. Subcommittee on Commerce, Justice, State, and Judiciary. Senate Dirksen Office Bldg., Washington, DC 20510. Voice: 202 224 7277. Legislation to appropriate funds for FCC, Board of International Broadcasting, NTIA, and U.S. Information Agency.

0750   Senate. Committee on Commerce, Science, and Transportation. Subcommittee on Communications. SHOB. Voice: 202 224 9340. Primary Senate center for legislation related to interstate and foreign communications (television, cable, local and long-distance telephone services, radio, international satellite communications). Oversees FCC and NTIA.

0751   Senate. Committee on the Judiciary. Subcommittee on Antitrust, Monopolies, and Business Rights. SHOB. Voice: 202 224 5710. Legislation related to anticompetitive practices and monopolies in communications, including cable and telephone industry practices.

0752   Senate. Committee on the Judiciary. Subcommittee on Patents, Copyrights, and Trademarks. SHOB. Voice: 202 224 8178.

0753   Senate. Committee on the Judiciary. Subcommittee on Technology and the Law. SHOB. Voice: 202 224 3406. Jurisdiction over Freedom of Information Act.

6-B-3-ii. Selected Congressional Agencies

0754   Board for International Broadcasting. Suite 400, 1201 Connecticut Ave., Washington DC 20036. Voice: 202 254 8040. Fax: 202 254 3929. Operated RFE and RL ca. 1975–1994. (To be folded into USIA in 1995.)

0755   Congressional Space Caucus. RHOB. Voice: 202 225 2631.

0756   Congressional Budget Office. Second and D Sts. SW, Washington DC 20515. Voice: 202 226 2621. Provides Congress with assessments of the economic impact of the federal budget.

0757   General Accounting Office. General Accounting Office Bldg., 441 G St., Washington DC 20548. Voice: 202 275 5067. Publications and reports: P.O. Box 6015, Gaithersburg MD 20884-6015. Voice: 202 512 6000. Investigative arm of Congress, independently auditing government agencies. Examines all matters related to receipt and disbursement of public funds; see its BIBLIOGRAPHY OF DOCUMENTS ISSUED BY THE GAO ON MATTERS RELATED TO ADP, IRM & TELECOMMUNICATIONS (Washington: GAO, 1983–date, annual).

0758   Office of Management and Communications. Voice: 202 512 6623. Fax: 202 512 6742.

0759   Division of Information Management and Technology. Voice: 202 512 6410. Fax: 202 512 6451. Government agency use of IRM.

0760   Library of Congress. 10 First St. SE, Washington DC 20540. Voice: 202 707 5000. Fax: 202 707 5844.

        1 **Congressional Research Service.** Science Policy Research Division. Voice: 202 707 9547. Provides objective, nonpartisan research, analysis, and information support to assist Congressional functions.

        2 **Copyright Office.** Voice: 202 707 8350. Issues copyrights and publishes a large number of regularly updated information leaflets on all aspects of copyright process.

0761   Office of Technology Assessment. 600 Pennsylvania Ave. SE, Washington DC 20003-8025. Personnel locator: 202 224 8713. Congressional and Public Affairs: 202 224 9241. Press: 202 228 6204. Publications: 202 224 8996. Fax: 202 228 6098. Internet: ota.gov. World-Wide Web: http://

www.ota.gov. Reports to Congress on the scientific and technical impact of government policies and proposed legislative initiatives. Involved in standards development process. Includes **Science, Information, and Natural Resources Division**. Telecommunication and Computing Technologies Program. Voice: 202 228 6760.

6-B-4. Resources on the Executive Branch

0762 **CONGRESSIONAL QUARTERLY'S GUIDE TO THE PRESIDENCY**. Washington: Congressional Quarterly, 1989, 1521 pp. Standard guide to history, organization, and operations of Presidency and executive departments, with bibliographies. Subject index.

0763 **FEDERAL REGULATORY DIRECTORY**. Washington: Congressional Information Service, 1979–date, annual. Narrative descriptions of executive and independent departments and agencies cover history, personnel and offices (with addresses, telephone numbers, etc.), organizational structure, libraries and other information resources, with bibliographies. Useful for solid overviews of FCC, FTC, etc.

0764 **FEDERAL STAFF DIRECTORY**. Mount Vernon, VA: Staff Directories, 1982–date, annual. Guide to executive offices and agencies and independent agencies. Keyword subject and name indexes. Useful for current personnel and telephone information for FCC and other offices. CD-ROM version STAFF DIRECTORIES ON CD-ROM available from Staff Directories; cumulates CONGRESSIONAL STAFF DIRECTORY (0742), JUDICIAL STAFF DIRECTORY (0784), and FEDERAL STAFF DIRECTORY: covers current editions, updated semiannually.

0765 Goehlert, Robert, and Hugh Reynolds. **THE EXECUTIVE BRANCH OF THE U.S. GOVERNMENT: A BIBLIOGRAPHY**. New York: Greenwood, 1989, 380 pp. Unannotated guide to literature on "history, development, organization, procedures, rulings, and policy" of Cabinet offices. Author and subject indexes.

6-B-5. Executive Branch

6-B-5-i. Selected Executive Offices

0766 **Office of Science and Technology Policy**. Executive Office Bldg., Washington DC 20500. Voice: 202 456 7116.

0767 **Office of U.S. Trade Representative**. Winder Bldg., 600 17th St., Washington DC 20506. Voice: 202 395 5797. Assistant representatives for GATT affairs and science and technology.

0768 **Office of Management and Budget**. Voice: 202 395 3914. Issues annual reports on telecommunications requirements of federal government; regular supervision of executive branch annual budget.

6-B-5-ii. Selected Cabinet Departments

0769 **Department of Agriculture. Rural Electrification Administration. Telephone Program**. 14th St. and Independence Ave., SW, Washington, DC 20250. Voice: 202 720 9540; 202 382 9540. Fax: 202 720 1725. World-Wide Web:http://fie.com/web/fed/agr. Makes loans and loan guarantees to rural electric and telephone utilities that serve rural areas. Administers Rural Telephone Bank that provides supplemental funding from nonfederal sources for telephone systems.

0770 **Department of Commerce**. Herbert C. Hoover Bldg., 14th St. between Pennsylvania and Constitution Aves. NW, Washington, DC 20230. Voice: 202 377 2000. Internet: gopher://gopher.esa.doc.gov.

  1 **Director of Information Resources Management** Voice: 202 482 1300.
  • Office of Information Policy and Technology. Voice: 202 482 3201.
  2 **Technology and Policy Analysis**. Voice: 202 482 4188.
  • Information Systems Technology. Voice: 202 482 0706.
  • Electronic Components and Information Technology. Voice: 202 482 1641.

3 **Office of Microelectronics and Instrumentation.** Voice: 202 482 1333.
- Telecommunications. Voice: 202 482 4466.

4 **International Trade Administration.** Under Secretary for International Trade. Voice: 202 377 3808. Promotes world trade and strengthening U.S. international trade and investment position.

5 **National Institute of Standards and Technology** (NIST). Technology Bldg. 225, Gaithersburg, MD 20899. Voice: 301 975 2816. Fax: 301 948 1784. Formerly National Bureau of Standards. Participates in standards development and promotes standards via publication, dissemination, and other information services. NIST also oversees National Information Standards Organization (NISO) that develops standards for publishing, libraries, information science. NIST offices include:
- Electronics and Electrical Engineering Laboratory. Voice: 301 975 2220.
- National Computer Systems Laboratory (NCSL). Voice: 301 975 2822. Manages Federal Information Processing Standards (FIPS) for Federal Automatic Data Processing Community (developed by NIST) and develops standards and guidelines (FIPS Standard Publications, or FIPS PUBS) coordinated with industry standards.
- National Center for Standards and Certification Information (NCSCI). Voice: 301 975 4040. Fax: 301 975 1559. Maintains a comprehensive collection of U.S. association, non-U.S. national, and international standards. Will not lend nor photocopy, but can help to obtain standards.

6 **National Technical Information Service** (NTIS). 5285 Port Royal Road, Springfield, VA 22161. Voice: 800 336 4700. Fax: 703 321 8547; 800 336 4700. Primary federal depository and distributor for research studies done for federal agencies. Supplies CCITT, ANSI, and other standards adopted by U.S. government agencies. Issues a variety of finding aids based on NTIS and other databases, including SATELLITE COMMUNICATIONS: VERY SMALL APERTURE TERMINALS (VSATs), 1985–1990 (1990); SPREAD SPECTRUM COMMUNICATIONS, 1970–1989 (1989) and 1964–1978 (1978); INFORMATION AND COMMUNICATION THEORY, 1981–1986 (1987); INTELLIGENT BUILDINGS: SHARED MULTITENANT TELECOMMUNICATIONS SERVICES, 1983–1987 (1987); TELECOMMUNICATION: ECONOMIC STUDIES, 1979–1985 (1985); FACSIMILE COMMUNICATION, 1970–1982 (1982); ELECTRONIC MAIL, 1977–1982 (1982); TELECOMMUNICATIONS IN MEDICINE, 1964–1980 (1980); TELECONFERENCING, 1964–1980 (1980); MICROWAVE COMMUNICATIONS, 1976–1979 (1979).

7 **National Telecommunications and Information Administration** (NTIA). Voice: 202 482 1551. Fax: 202 482 1635. Develops telecommunication and information policy for executive branch. Develops and presents U.S. plans and policies in international fora and coordinates U.S. telecommunications and information policy positions in consulation with FCC, Department of State, and other agencies. Manages federal use of radio spectrum. Government base for National Information Infrastructure (NII) programs. Conducts research on telecommunication. Provides information to federal and state agencies. Awards construction grants to noncommercial telecommunication services. Administers National Endowment for Children's Educational Television. Issues PUBLICATIONS ABSTRACTS (1979–date) through NTIS, an annual list of NTIA staff publications on results of research and engineering. NTIA includes these major units:
- Policy Coordination and Management. Voice: 202 482 1835
- Office of Policy Analysis and Development. Voice: 202 377 1880. Domestic policy development.
- Office of Spectrum Management. Voice: 202 377 1850.
- Interdepartment Radio Advisory Committee. (IRAC). Voice: 202 377 0599. Oldest federal telecommunications agency (created in 1922), this is a frequency spectrum coordination body for all federal users.
- Frequency Management Advisory Council. Voice: 202 377 1850.
- Office of International Affairs. Voice: 202 482 1304
- Information Infrastructure Task Force. Voice: 202 482 1840. World-Wide Web: iitf.doc.gov.Issued THE NATIONAL INFORMATION INFRASTRUCTURE: AGENDA FOR ACTION (1993) and all subsequent NII analyses.

0771 **Department of Defense.** Pentagon, Washington DC 20318-0001. Voice: 703 697 9121. World-Wide Web: http:// enterprise.osd.mil.

1 **Director for Command, Control, and Communications Systems,** J-6. Voice: 703 695 6478.

2 **Defense Information Systems Agency.** Voice: 703 692 9012; 703 692 0018; 703 692 0016. Sometimes called National Communications System. Responsible for planning, developing, and supporting command, control, communications, and information systems that serve the needs of the National Command Authorities under all conditions of peace and war. Provides communications support for Federal Emergency Management Agency.

3 **National Security Agency/Central Security Service.** Voice: 301 688 6311. Supports U.S. government activities to protect U.S. communications and produce foreign national intelligence. Responsible for signals intelligence and communications security activities of government. Operates National Computer Security Center.

4 **Advanced Research Projects Agency.** Voice: 202 545 6700; 703 696 2444.

5 **Joint Service School. Information Resources Management College.** Voice: 703 433 3000,

6 **Naval Computer and Telecommunications Command.** Voice: 703 282 0550. Fax: 703 282 0366.

7 **Deputy Assistant Secretary of the Air Force for Communications, Computers, and Logistics.** Voice: 703 697 3624. Fax: 703 693 7553.

8 **Deputy Chief of Staff, U.S. Air Force, for Command, Control, Communications, and Computers.** Voice: 703 695 6324.

0772 **Department of Justice.** Main Justice Bldg., 10th and Constitution Ave., Washington DC 20530. Locator 202 514 2000. World-Wide Web: http://justice2.usdoj.gov.

1 **Assistant Attorney General for Policy Development. Information and Privacy.** Rm. 7238. Voice: 202 514 4251.

2 **Executive Office for U.S. Attorneys. Assistant Director for Telecommunications and Technology Development.** Patrick Henry Bldg., 601 D St., Washington DC 20004. Voice: 202 501 6924.

3 **Antitrust Division.** Assistant Attorney General. 555 4th St., NW, Washington, DC 20001. Voice: 202 514 5621; 202 633 2401. Investigates and litigates antitrust cases in communications. Participates in agency proceedings and rulemaking concerning communications; monitors and analyses legislation. Oversees continuing matters concerning AT&T divestiture.

4 **Regulatory Affairs.** Deputy Assistant Attorney General. Voice: 202 633 2404. Section on Communications and Finance. Voice: 202 272 4247.

0773 **Department of State.** Main State Bldg., Washington, DC 20520. Locator: 202 647 4000.

1 **Bureau of Economic and Business Affairs.** Voice: 202 647 7971. Handles trade-related international telecommunication issues.

2 **Bureau of International Communications and Information Policy.** Voice: 202 647 5727; 202 647 5832. Fax: 202 647 5957. Develops and manages international communication information policy for State Deptartment. Liaison for other federal agencies and departments and the private sector in international communications issues.

• International Telecommunications Advisory Committee (formerly U.S. National Committee for CCITT.) Voice: 202 647 2592. Fax: 202 647 7407. Represents U.S. interests in ITU-T sector (formerly CCITT), advising and reporting back to State Department; composed of study groups for telecommunication services and policy; switching, signaling, and ISDN; telephone network operations; and data network and telematic terminals that reflect CCITT activities; membership includes organizations, product and service providers.

3 **Office of Radio Spectrum Policy.** Voice: 202 647 2592.

4 **Office of Telecommunications and Information Standards.** Voice: 202 647 5230.

6-B-5-iii. Selected Independent Agencies (except FCC)

0774   **Federal Trade Commission. Bureau of Competition**. Sixth and Pennsylvania Ave., Washington, DC 20580. Voice: 202 326 2556. Shares anti-trust oversight and enforcement with the Department of Justice.

0775   **General Services Administration**. General Services Bldg., 18th and F Sts., Washington DC 20405.
    1 Information Resources Management Service. Voice: 202 501 1000. Fax: 202 501 0022. Purchases and leases automatic data processing and telecommunication equipment for federal government. Advises agencies and administers implementation of systems. Operates Federal Telecommunications Services networks.
    2 Information Security Oversight Office. Associate Director for Policy. Voice: 202 634 6150.
    3 Office of Federal Telecommunications System (FTS) 2000. Voice: 202 208 7493. Provides common-user, long distance telecommunications services (FTS-2000). Provides leadership, policy, program direction, and program oversight for a timely, high-quality, innovative, and cost-effective government-wide program for intercity telecom-munications services.

0776   **National Aeronautics and Space Administration**. 400 Maryland Ave. SW, Washington DC 20546. Voice: 202 453 8400. World-Wide Web: http://hypatia.gsfc.gov.
    1 Office of Policy Coordination and International Relations. Voice: 202 358 0400.
    2 Office of Space Communications. Voice: 202 453 2800.

0777   **U.S. International Trade Commission**. 500 E St. SW, Washington DC 20436. Voice: 202 205 2000. Fax: 202 205 2798.
    1 Division of Service and Electronic Technology. Voice: 202 205 3440.
    2 Office of International Competitiveness. Voice: 202 205 3124.
    3 Tariff Affairs and Trade Agreements. Voice: 202 205 2592.
    4 Unfair Import Investigations. Voice: 202 205 2561.

6-b-5-iv Federal Communications Commission

6-B-5-iv-a. Guides to the FCC

0778   Federal Communications Commission, Public Service Division. **INFORMATION SEEKERS GUIDE: HOW TO FIND INFORMATION AT THE FCC**. Washington: FCC, October 1994 (regularly revised), 40 pp. Unfortunately this latest version does not include the dramatic FCC reorganization of 1994. Still, it is a useful "key" to finding people, documents, and processes at the commission.

0779   _____. Office of the Managing Director. **FCC TELEPHONE DIRECTORY**. Washington: FCC, continually revised. Once issued four times a year, this loose-leaf directory is now revised continually, a section at a time. Most useful is the "organizational listing" (blue pages) which detail FCC structure and key personnel. As of December 1994, this includes the new bureaus created in 1994's reorganization.

0780   _____. **ANNUAL REPORT**. Washington: GPO, 1935–date, annual (editions for 1942–44 originally published only in limited mimeograph version; 1935–55 issues reprinted by Arno Press in 3 vols., 1971). Annual report on FCC regulation of interstate and foreign communications. Includes FCC operating data: employees, funding, status of cases. Regulatory activities data: authorizations, licensing, permits, applications, denials, revocations, by class of operation; Field Operations Bureau statistics for applications, examinations, licensed operators, by type; private radio statistics for authorized stations, transmitters, authorization requests, by type; equipment authorization applications, acceptances, notifications, certifications, changes, by type. Data from reports by regulated companies and stations and FCC files. Indexed in STATISTICAL MASTERFILE (0513).

0781   Hilliard, Robert. **THE FEDERAL COMMUNICATIONS COMMISSION: A PRIMER**. Stoneham, MA: Focal Press, 1991, 115 pp. A brief survey of the FCC and its functions, emphasizing mass communications.

6-B-5-iv-b. FCC and Its Major Divisions

0782   **Federal Communications Commission**, 1919 M. St. NW, Washington, DC 20554. Consumer Assistance Branch voice: 202 418 0190. World-Wide Web: http://www.fcc.gov. [Note: FCC is scheduled to move to the Portals building complex in the southwest part of Washington, beginning in late 1995.] Agency holds primary responsibility for telecommunications authorization and regulation. Participates in development of technical standards, and oversees operator compliance. Publishes, as formal legal record of all proceedings, FCC RECORD (1986-date, biweekly); formerly FCC REPORTS (1934-1986). Full text is available online on LEXIS/NEXIS (0728).

1 **Office of the Chairman**. Voice: 202 418 1000.

2 **Office of Public Affairs**. Voice: 202 418 0500. Includes FCC Library, Room 639, Voice: 202 418 0450.

3 **Office of Legislative and Inter-Governmental Affairs**. Voice: 202 418 1900

4 **Office of Plans and Policy**. Voice: 202 418 2030. "Think tank" and research arm (largely economic in approach) for the FCC.

5 **Office of Engineering and Technology**. Voice: 202 739 0700. Includes divisions on spectrum engineering, propagation analysis, engineering evaluation, equipment authorization, sampling and measurements, and technical standards.

6 **Cable Services Bureau**. 2033 M St. NW, Washington, DC 20554. Voice: 202 416 0856. The FCC had a cable bureau in the 1970s which disappeared with deregulation. This new version, set up in 1993, grew out of Congressional re-regulation of cable. Includes divisions on consumer protection, policy and rules, and technical services.

7 **Common Carrier Bureau**. Voice: 202 418 1500. Focus of telecommunications regulation, this unit's role has been streamlined with both deregulation and formation of newer bureaus. Includes divisions on accounting and audits (2000 L St. NW), domestic facilities (2025 M St. NW), enforcement (2025 M St. NW), industry analysis (1250 23rd St. NW), tariffs, and policy and program planning.

8 **Compliance and Information Bureau**. Voice: 202 418 1100. Formerly the Field Operations Bureau, this unit, renamed in 1994, operates FCC field offices across the country and includes divisions on enforcement and engineering as well as the emergency alert system (formerly emergency broadcast system).

9 **International Bureau**. 2000 M St. NW, Washington, DC 20554 Voice: 202 418 0420. Grew out of a smaller office to become a bureau in 1994 and includes all FCC activities interacting with other countries and the ITU. Includes divisions on telecommunications [which centralizes all FCC satellite functions, domestic and international], and planning and negotiations (2025 M St. NW).

10 **Mass Media Bureau**. Voice: 202 418 2780. Centralizes all radio and television authorization and regulation in divisions on audio services, enforcement, (2025 M St. NW), Policy and rules (2025 M St. NW), and video services (which includes MMDS services, 2033 M St. NW).

11 **Wireless Telecommunications Bureau**, 2025 M St. NW, Washington, DC 20554. Voice: 202 418 0600. Formed out of the former Private Radio Bureau in 1994 to centralize all domestic wireless regulation (except satellite services), this includes divisions on auctions (of spectrum for selected services), commercial (mobile) radio, enforcement, policy, private (mobile) radio, operations (located in Gettysburg, PA), and licensing.

6-B-6. Resources on the Federal Court System

0783   **CONGRESSIONAL QUARTERLY'S GUIDE TO THE U.S. SUPREME COURT**. Washington: Congressional Quarterly, 1990 (2nd ed.), 1060 pp. Standard guide to history and practices of Supreme Court, with indexes of subjects, cases, lists of major decisions, and bibliographies.

0784   **JUDICIAL STAFF DIRECTORY**. Mount Vernon, VA: Staff Directories, 1986–date, annual. Guide to federal judiciary and staffs, with biographies. Keyword subject and name indexes. CD-ROM version STAFF DIRECTORIES ON CD-ROM available from Staff Directories; cumulates CONGRESSIONAL STAFF DIRECTORY (0742) and FEDERAL STAFF DIRECTORY (0764): covers current editions, updated semiannually.

0785   **PIKE AND FISCHER'S DESK GUIDE TO COMMUNICATIONS LAW RESEARCH**. Robert Emeritz and others, editors. Bethesda, MD: Pike & Fischer, 1993–date, to be revised biennially. Essentially a guide to accessing information published by FCC. Detailed explanation of federal telecommunications law, overview (with diagrams) of FCC structure, directory of FCC headquarters and field offices (with telephone numbers), and copies of principal FCC forms, in addition to explanations on using PIKE & FISCHER RADIO REGULATION and other products. Glossary, bibliographies.

   6-B-6-i. Federal Court System

0786   **The Supreme Court of the United States**, U.S. Supreme Court Building, 1 First St., NE, Washington, DC 20543. Voice: 202 479 3211 (Public Information Office). Decides some cases appealed from lower federal courts but as with other topic areas, rejects many such appeals, allowing lower court decision to stand. There are many ways to find texts of decisions (most include the abbreviation used in legal citations):

   1 **UNITED STATES REPORTS (U.S.)** (many private reporters and publishers prior to 1922; Washington, DC: GPO, 1922–date). Official version of all Supreme Court decisions, this series now runs to nearly 500 volumes and should be found in any good law school library. First to appear during a session are slip opinions, followed by preliminary prints and then final bound volumes (3–4 per term). Though the official final authority on wording, volumes appear as much as three years late, making following private sources invaluable.

   2 **SUPREME COURT REPORTER (S.Ct.)** Minneapolis: West Publishing Co., 1882–date. Perhaps the best and most widely cited, though unofficial, source for Supreme Court decisions—best because it publishes decisions far faster than the official source. Decisions appear first in biweekly advance sheets, followed by interim volumes and then final volumes, usually three per term.

   3 **UNITED STATES SUPREME COURT REPORTS: THE LAWYER'S EDITION** Rochester, NY: The Lawyers Cooperative Publishing Company, 1790–1955 First Series, 100 vols; 1956–date Second Series. Another commercial (unofficial) resource which includes summaries of lawyers' written briefs.

   4 **UNITED STATES LAW WEEK** Washington, DC: Bureau of National Affairs, 1933–date, weekly. Publishes Supreme Court decisions within about a week of their issue. These are cumulated in a loose-leaf service.

   5 **UNITED STATES SUPREME COURT BULLETIN** Chicago: Commerce Clearing House, 1980–date. Also publishes Supreme Court decisions within a week or so of their issue, cumulating in a loose-leaf notebook.

   6 **LANDMARK BRIEFS AND ARGUMENTS OF THE SUPREME COURT OF THE UNITED STATES** Frederick, later Bethesda, MD: University Publications of America, 1978–date. Prints selected briefs and arguments of Supreme Court cases.

0787   **United States Court of Appeals for the District of Columbia Circuit**, U.S. Court House, Third and Constitution Ave., NW, Washington, D.C. 20001. Voice: 202 535 3308. This is but one of thirteen such courts, important to telecommunications as the "expert court" dealing with most appeals from FCC decisions. Its decisions are found in:

   1 **FEDERAL REPORTER (F., F.2d, F.3d)** Minneapolis: West Publishing, 1880–1924 for First Series (300 volumes); 1925–1993 for Second Series (999 volumes); and 1993–date for Third Series. The best printed commercial resource for decisions of all the federal appeals courts. Note: prior to 1932, this series included decisions of U.S. District Courts.

0788   **United States District Court for the District of Columbia,** U.S. District Court House, Third St. at Constitution Ave. NW, Washington, D.C. 20001. Voice: 202 535 3522 (Clerk's Office, Room 1825A, First Floor). This is one of 89 federal district courts (1–4 per state), and is especially important to telecommunications as the "home" of the continuing proceedings concerning the AT&T divestiture and its aftermath under the supervision of Judge Harold H. Greene. The reference room is open from 9:00 am to 4:30 pm weekdays. AT&T case documents are found there under one of three separate "docket" or proceeding numbers: 74–1698 (1974–84), 82–0025 (1982–date), and 82–0192 (also 1982–date), each of which is chronologically indexed by a Docket Sheet (each "sheet" is actually more than 100 pages long). Other sources for texts of decisions include:

1 **FEDERAL SUPPLEMENT (F.Supp.)** Minneapolis: West Publishing, 1923–date. The commercial resource for all federal district courts, including this one, though it includes only a small proportion of all cases decided (the judge writing the opinion make the decision on whether to publish).

2 **FEDERAL RULES DECISIONS (F.R.D.)** Minneapolis: West Publishing, 1941–date. Includes those U.S. District Court decisions involving federal rules of civil procedure, and thus may include some communication-related decisions based on FCC rules and the like.

6-C. State Governments

6-C-1. Resources for State Governments

0789   **PROFILES OF REGULATORY AGENCIES OF THE UNITED STATES AND CANADA.** Washington: National Association of Regulatory Utility Commissioners, 1992–date, annual. Provides directory information, histories, practices and procedures, and comparative data for federal, state, and provincial regulatory agencies. Expands information formerly included in NARUC ANNUAL REPORT ON UTILITY AND CARRIER REGULATION (1973–1991); companion volume to UTILITY REGULATORY POLICY IN THE UNITED STATES AND CANADA (1992–date).

0790   **STATE ADMINISTRATIVE OFFICIALS CLASSIFIED BY FUNCTION.** Lexington, KY: Council of State Governments, 1961–date, annual. Title has varied. Classified subject listings of specialized offices in each state. Relevant listings under "Telecommunications," "Public Utility Regulation," "Aeronautics," etc.

0791   **TELECOMMUNICATIONS.** Lansing: Legislative Service Bureau Library, 1993, 4 leaves. (Michigan Legislative Topics). Subtitle: "A bibliography compiled for use of the Michigan legislature." Identifies about 50 reports and studies issued by state and multi-state offices on local and regional impacts of telecommunications and its regulation.

6-C-2. States Governments and Offices

6-C-2-i. State Public Utilities Commissions

0792   **Alabama Public Service Commission.** P.O. Box 991, Montgomery AL 36102. Voice: 205 242 5209. Fax: 205 240 3079. Jurisdiction over radio common carriers, COCOTs; resellers; telephone; telegraph. Maintains library; focuses on telecommunications. Publishes ANNUAL REPORT (1881–date); early volumes annual and biennial reports of the Railroad Commission.

0793   **Alaska Public Utilities Commission.** Commerce and Economic Development Department, 420 L. St., Suite 100, Anchorage AK 99501. Voice: 907 276 6222. Fax: 907 276 0160. Jurisdiction over telecommunications, cable television; radio common carriers. Maintains library; focuses on telecommunications. Publishes ANNUAL REPORT (1970–date); indexed in STATISTICAL MASTERFILE (0513).

0794   **Arizona Corporation Commission.** 1200 W. Washington St., Phoenix, AZ 85007. Voice: 602 542 3076. Fax: 602 542 4870. Jurisdiction over telephone; cellular; telegraph. Maintains library. Publishes ANNUAL REPORT (1913–date); some early volumes triennial.

0795   **Arkansas Public Service Commission**. 1000 Center Bldg., P.O. Box C-400, Little Rock, AR 72203. Voice: 501 682 1794. Fax: 501 682 5731. Jurisdiction over telephone. Maintains library. Publishes ANNUAL REPORT (1899–date); early volumes annual and biennial reports of the Railroad Commission; indexed in STATISTICAL MASTERFILE (0513).

0796   **California Public Utilities Commission**. California State Bldg., 505 Van Ness Ave., San Francisco, CA 94102 3298. Voice: 415 557 0647. Fax: 415 557 1923. Jurisdiction over telephone; radio common carriers; telegraph. Maintains library. Publishes ANNUAL REPORT (1880–date); indexed in STATISTICAL MASTERFILE (0513). Also publishes INTEREXCHANGE CARRIER (1991–date), LOCAL EXCHANGE CARRIER (1991–date), CELLULAR WHOLESALE (1992–date), CELLULAR RESELLER (1991–date).

0797   **Colorado Public Utilities Commission**. Logan Tower, Office Level 2, 1580 Logan St., Room 203, Denver, CO 80203. Voice: 303 894 2070. Jurisdiction over telephone; telegraph. Maintains library. Publishes ANNUAL REPORT (1914–date); formerly BIENNIAL REPORT (1911–1913). Also publishes TELECOMMUNICATIONS UTILITIES REGULATED BY THE COLORADO PUBLIC UTILITIES COMMISSION (1993–date).

0798   **Connecticut Department of Public Utility Control**. One Central Park Plaza, New Britain, CT 06051. Voice: 203 827 1553. Fax: 203 827 2613. Jurisdiction over telephone; cable television; telegraph. Maintains library; focuses on telecommunications. Publishes REPORT (1979–date, annual). Also publishes ANNUAL REPORT TO THE GENERAL ASSEMBLY ON THE STATUS OF TELECOMMUNICATIONS IN CONNECTICUT (1987–date); and DPUC REVIEW OF TELECOMMUNICATIONS POLICIES (1993).

0799   **Delaware Public Service Commission**. 1560 S. DuPont Hwy, Dover, DE 19903 0457. Voice: 302 736 4247. Fax: 302 736 4849. Jurisdiction over telephone; cable television. Publishes REPORT (1950–date, biennial).

0800   **District of Columbia Public Service Commission**. 450 Fifth St. NW, Washington, DC 20001. Voice: 202 626 5100. Fax: 202 638 2330. Jurisdiction over telephone; telegraph. Maintains library. Publishes ANNUAL REPORT (1913–date); indexed in STATISTICAL MASTERFILE (0513).

0801   **Florida Public Service Commission**. 101 E. Gaines St., Tallahassee, FL 32399 0850. Voice: 904 488 4733. Fax: 904 487 0509. Jurisdiction over telephone (LEC, IXC, STS and payphones). Maintains library; focuses on telecommunications. Publishes ANNUAL REPORT (1888–date); early volumes REPORT OF THE FLORIDA RAILROAD AND PUBLIC UTILITIES COMMISSION and variant titles; indexed in STATISTICAL MASTERFILE (0513).

0802   **Georgia Public Service Commission**. 244 Washington St., SW, Atlanta, GA 30334. Voice: 404 656 7491. Fax: 404 487 2341. Jurisdiction over telephone; radio common carriers; telegraph. Publishes REPORT (1894–date, annual); early volumes REPORT OF THE RAILROAD COMMISSION.

0803   **Hawaii Public Utilities Commission**, Department of Budget & Finance, 465 S. King St., Honolulu, HI 96813. Voice: 808 548 3990. Fax: 808 548 4376. Jurisdiction over telephone; radio common carriers. Publishes ANNUAL REPORT (1913–date).

0804   **Idaho Public Utilities Commission**. Statehouse, 472 W. Washington St., Boise, ID 83720. Voice: 208 334 0300. Fax: 208 334 3762. Jurisdiction over telephone. Publishes REPORT (1914–date, annual); indexed in STATISTICAL MASTERFILE (0513).

0805   **Illinois Commerce Commission**. Leland Bldg., 527 East Capitol Ave., Springfield, IL 62794-9280. Voice: 317 524 5054. Fax: 217 782 1042. Jurisdiction over telecommunications carriers. Maintains library. Publishes ANNUAL REPORT (1921–date). Also publishes ANNUAL REPORT ON TELECOMMUNICATIONS 1985–date). Also publishes staff papers on telecommunications issues of local interest.

0806   **Indiana Utility Regulatory Commission**. 913 State Office Bldg., Indianapolis, IN 46204. Voice: 317 232 2701. Fax: 317 232 6758. Jurisdiction over telephone; telegraph. Publishes ANNUAL REPORT (1914–date); early volumes REPORT OF THE PUBLIC SERVICES COMMISSION.

0807   **Iowa State Utilities Board**. Lucas State Office Bldg., Des Moines, IA 50319. Voice: 515 281 5979. Fax: 515 281 5329. Jurisdiction over telephone (rates for over 15,000 customers or access lines); telegraph. Maintains library; focuses on telecommunications. Publishes ANNUAL REPORT (1878– date); early volumes report of Board of Railroad Commissioners and State Commerce Commission.

0808   **Kansas Corporation Commission**. Docking State Office Bldg., Topeka, KS 66612-1571. Voice: 913 296 3355. Fax: 913 296 3596. Jurisdiction over telephone; radio common carriers; telegraph. Maintains library. Publishes REPORT (1911–date, biennial).

0809   **Kentucky Public Service Commission**. 730 Schenkel Lane, P.O. Box 615, Frankfort, KY 40602. Voice: 502 836 3940. Fax: 502 836 7279. Jurisdiction over cooperative electric and telephone utilities; telephone (private); radio common carriers. Maintains library. Publishes BIENNIAL REPORT (1935–date); early volumes annual; indexed in STATISTICAL MASTERFILE (0513).

0810   **Louisiana Public Service Commission**. Department of Public Service, One American Place, Suite 1630, Baton Rouge, LA 70825-1697. Voice: 504 342 4427. Jurisdiction over telephone; radio common carriers; telegraph. Publishes ANNUAL (1900–date); early volumes reports of the Railroad Commission.

0811   **Maine Public Service Commission**. 242 State St., State House Station 18, Augusta, ME 04333. Voice: 207 289 3831. Fax: 207 289 1039. Jurisdiction over telephone; radio common carriers; telegraph. Maintains library. Publishes ANNUAL REPORT (1915–date).

0812   **Maryland Public Utilities Commission**. American Bldg., 231 E. Baltimore St., Baltimore, MD 21202-3486. Voice: 301 333 6000. Fax: 301 333 6495. Jurisdiction over telephone. Maintains library. Publishes REPORT (1910–date, annual).

0813   **Massachusetts Department of Public Utilities**. 100 Cambridge St., Boston, MA 02202. Voice: 617 727 3500. Fax: 617 723 8812. Jurisdiction over telephone; radio common carriers; telegraph. Maintains library. Publishes ANNUAL REPORT (1914–date).

0814   **Michigan Public Service Commission**. Mercantile Bldg., 6545 Mercantile Way., P.O. Box 30221, Lansing, MI 48909. Voice: 517 334 6422. Fax: 517 882 5170. Jurisdiction over telephone. Maintains library. Publishes ANNUAL REPORT (1919–date).

0815   **Minnesota Public Utilities Commission**. 780 American Center Bldg., 150 E. Kellogg Blvd., St. Paul, MN 55101. Voice: 612 696 7124. Fax: 612 297 1959. Jurisdiction over telephone; telegraph. Maintains library; focuses on telecommunications. Publishes BIENIAL REPORT (1887–date); early volumes annual reports of the Railroad and Warehouse Commission.

0816   **Mississippi Public Service Commission**. Walter Sillers State Office Bldg., P.O. Box 1174, Jackson, MS 39215-1174. Voice: 601 961 5400. Fax: 601 297 5469. Jurisdiction over telephone; radio common carriers; telegraph. Maintains library; focuses on telecommunications. Publishes ANNUAL REPORT (1939–date); early volumes biennial.

0817   **Missouri Public Service Commission**. Truman State Office Bldg., P.O. Box 360, Jefferson City, MO 65102. Voice: 314 751 3234. Fax: 314 751 1847. Jurisdiction over telecommunications. Publishes ANNUAL REPORT (1913–date); early variant titles.

0818   **Montana Public Service Commission**. 2710 Prospect Ave., Helena, MT 59620. Voice: 406 444 6169. Fax: 406 444 7618. Jurisdiction over telephone; radio common carriers; telegraph. Maintains library; focuses on telecommunications. Publishes ANNUAL REPORT (1908–date); early volumes reports of the Railroad Commission.

0819   **Nebraska Public Service Commission**. 300 The Atrium, 1200 N St., P.O. Box 94927, Lincoln, NE 68509-4927. Voice: 402 471 3101. Fax: 402 471 0254. Jurisdiction over telephone.

Maintains library; focuses on telecommunications. Publishes BIENNIAL REPORT (1908–date); annual report through 1972. Also publishes ANNUAL REPORT TO THE LEGISLATURE ON THE STATUS OF THE NEBRASKA TELECOMMUNICATIONS INDUSTRY (1988–date, annual).

0820   **Nevada Public Service Commission**. 727 Fairview Dr., Carson City, NV 89710. Voice: 702 687 6001. Fax: 702 687 6110. Jurisdiction over telephone; radio common carriers; telegraph. Maintains library. Publishes COMBINED BIENNIAL REPORT (1907–date); early volumes separate reports of Railroad and Public Services commissions.

0821   **New Hampshire Public Utilities Commission**. 8 Old Suncook Rd., Bldg. No. 1, Concord, NH 03301-5185. Voice: 603 271 2431. Fax: 603 271 3878. Jurisdiction over telephone; telegraph. Maintains library. Publishes BIENNIAL REPORT (1911–date); volumes annual through 1972.

0822   **New Jersey Board of Public Utilities**. Two Gateway Center, Newark, NJ 07102. Voice: 201 648 2026. Fax: 201 648 2836. Jurisdiction over telephone; cable television; telegraph. Publishes ANNUAL REPORT (1913–date); formerly ANNUAL REPORT OF THE BOARD OF RAILROAD COMMISSIONERS (1907–1909).

0823   **New Mexico Public Service Commission**. Marian Hall, 224 E. Palace, P.O. Box 2205, Santa Fe, NM 87503-2205. Voice: 505 827 6940. Fax: 505 827 6973. Jurisdiction over telephone; radio common carriers; telegraph. Publishes ANNUAL REPORT (1941/1942–date).

0824   **New York Public Service Commission**. Empire State Plaza, Albany, NY 12223. Voice: 518 474 2510. Fax: 518 474 7146. Jurisdiction over telephone; cellular radio carriers. Maintains library. Publishes ANNUAL REPORT (1909–date).

0825   **North Carolina Utilities Commission**. Department of Commerce, 430 N. Salisbury St., Dobbs Bldg., P.O. Box 29510, Raleigh, NC 27626-0510. Voice: 919 733 4249. Fax: 919 733 7300. Jurisdiction over telephone; radio common carriers. Maintains library. Publishes REPORT (1899–date, biennial); some issues annual.

0826   **North Dakota Public Service Commission**. State Capitol, Bismarck, ND 58505. Voice: 701 224 2400. Fax: 701 224 2410. Jurisdiction over telephone; radio common carriers. Publishes ANNUAL REPORT (1910–date, biennial); formerly annual and biennial reports of the Commissioners of Railroads; indexed in STATISTICAL MASTERFILE (0513).

0827   **Ohio Public Utilities Commission**. 180 E. Broad St., Columbus, OH 43266-0573. Voice: 614 466 3016. Fax: 614 466 9546. Jurisdiction over telephone; radio common carriers; telegraph. Maintains library; focuses on telecommunications. Publishes REPORT (1913–date, annual).

0828   **Oklahoma Corporation Commission**. Jim Thorpe Office Bldg., Oklahoma City, OK 73105. Voice: 405 521 2261. Fax: 405 521 6045. Jurisdiction over telephone; telegraph. Maintains library. Publishes ANNUAL REPORT (1908–date).

0829   **Oregon Public Utility Commission**. 300 Labor and Industries Bldg., Salem, OR 97310. Voice: 503 378 5849. Fax: 503 373 7752. Jurisdiction over telephone. Maintains library. Publishes ANNUAL REPORT (1907–date).

0830   **Pennsylvania Public Utility Commission**. 104 N. Office Bldg., P.O. Box 3265, Harrisburg, PA 17120. Voice: 717 783 1740. Fax: 717 787 4193. Jurisdiction over telephone; radio common carriers; telegraph. Maintains library. Publishes ANNUAL REPORT (1914–date).

0831   **Puerto Rico Public Service Commission**. P.O. Box 870, Hato Rey Station, San Juan, PR 00919-0870. Voice: 809 751 5050. Fax: 809 753 9677. Jurisdiction over cable television.

0832   **Puerto Rico Telecommunications Regulatory Commission**. American Airlines Bldg., Second Floor, 1509 Lopez Landron St., San Juan, PR 00910. Mail address: GPO 5467, San Juan, PR 00919-5467. Voice: 809 268 4141.

0833   **Puerto Rico Telephone Authority**. P.O. Box 360998, Caparra Heights, PR 00936-0998. Voice: 809 782 8282.

0834　**Rhode Island Public Utilities Commission**. 100 Orange St., Providence, RI 02903. Voice: 401 277 3500. Fax: 401 277 6805. Jurisdiction over telephone; cable television; telegraph. Maintains library. Publishes ANNUAL REPORT (1912–date).

0835　**South Carolina Public Service Commission**. 111 Doctors Circle, P.O. Drawer 11649, Columbia, SC 29203. Voice: 803 737 5100. Fax: 803 737 5199. Jurisdiction over telephone; radio common carriers. Publishes ANNUAL REPORT (1879–date); early volumes reports of the Railroad Commissioner; indexed in STATISTICAL MASTERFILE (0513).

0836　**South Dakota Public Utilities Commission**. Capitol Bldg., Pierre, SD 57501. Voice: 605 773 3201. Fax: 605 773 3686. Jurisdiction over telephone; radio common carriers. Publishes ANNUAL REPORT (1890–date); early volumes annual and biennial reports of the Railroad Commissioners; indexed in STATISTICAL MASTERFILE (0513).

0837　**Tennessee Public Service Commission**. 460 James Robertson Pkwy., Nashville, TN 37319. Voice: 615 741 3668. Fax: 615 741 2336. Jurisdiction over telephone (co-op exempt); radio common carriers (not cellular). Publishes REPORT (1897–date, annual); early volumes reports of the Railroad Commission; indexed in STATISTICAL MASTERFILE (0513).

0838　**Texas Public Utility Commission**. 7800 Shoal Creek Blvd., Austin, TX 78757. Voice: 512 458 0100. Fax: 512 458 8340. Jurisdiction over telephone. Maintains library; focuses on telecommunications. Publishes ANNUAL REPORT (1976–date). Also publishes biennial report to legislature, STATUS OF COMPETITION IN LONG DISTANCE AND LOCAL TELECOMMUNICATIONS MARKETS IN TEXAS (1989–date), survey of telephone services, and financial reports.

0839　**Utah Public Service Commission**. 160 E. 300 South, P.O. Box 45585, Salt Lake City, UT 84145. Voice: 801 530 6716. Fax: 801 530 6796. Jurisdiction over telephone; radio common carriers; telegraph. Maintains library. Publishes REPORT (1917–date, annual).

0840　**Vermont Public Service Board**. 120 State St., State Office Bldg., Montpelier, VT 05602. Voice: 802 828 2358. Fax: 802 828 2342. Jurisdiction over telephone; cable television. Maintains library. Publishes BIENNIAL REPORT (1856–date); volumes through 1909 annual and biennial reports of the Board of Railroad Commissioners.

0841　**Virgin Islands Public Services Commission**. P.O. Box 40, Charlotte Amalie, St. Thomas, VI 00804. Voice: 809 776 1291. Fax: 809 774 4971. Jurisdiction over cable television.

0842　**Virginia State Corporation Board**. Jefferson Bldg., P.O. Box 1197, Richmond, VA 23209. Voice: 804 786 3608. Fax: 804 371 7376. Jurisdiction over telephone (investor); telephone and electric cooperatives; radio common carriers; cellular radio carriers. Maintains library. Publishes ANNUAL REPORT (1877–date); early volumes reports of the Railroad Commissioner.

0843　**Washington Utilities and Transportation Commission**. Chandler Plaza Bldg., 1300 S. Evergreen Park Dr., SW, Olympia, WA 98504-8002. Voice: 206 753 6423. Fax: 206 586 1150. Jurisdiction over telecommunications. Maintains library; focuses on telecommunications. Publishes REPORT (1950–date, biennial).

0844　**West Virginia Public Service Commission**. 201 Brooks St., P.O. Box 812, Charleston, WV 25323. Voice: 304 340 0300. Fax: 304 340 0325. Jurisdiction over telephone; radio common carriers; telegraph. Maintains library; focuses on telecommunications. Publishes ANNUAL REPORT (1914–date); some early volumes biennial.

0845　**Wisconsin Public Service Commission**. 477 Hill Farms State Office Bldg., P.O. Box 7854, Madison, WI 53707. Voice: 608 266 2001. Fax: 608 266 3957. Jurisdiction over telephone; radio common carriers. Maintains library. Publishes BIENNIAL REPORT (1906–date); early volumes reports of the Railroad Commission. Also publishes TELEPHONE OPERATING STATISTICS (1986–date).

0846　**Wyoming Public Service Commission**. 700 W. 21st St., Cheyenne, WY 82002. Voice: 307 777 7427. Fax: 307 777 5700. Jurisdiction over telephone; radio common carriers; other intrastate telecommunications companies; telegraph. Maintains library. Publishes ANNUAL REPORT (1916–date); volumes biennial through 1974.

6-C-2-ii. Multi-State Groups

0847   **Council of State Governments**. Iron Works Pike, P. O. Box 11910, Lexington, KY 40578-1901. Voice: 606 231 1939. Fax: 606 231 1858. Works to strengthen state government. Sponsors **National Association of State Telecommunications Directors**, voice: 606 231 1873, fax: 606 231 1928. Addresses national communications policies and regulatory issues. Publishes many reports and studies on impacts of telecommunications on states, including TELECOMMUNICATIONS AND STATE ECONOMIC DEVELOPMENT (1990); TELECOM-MUTING: TRANSPORTING IDEAS INSTEAD OF PEOPLE (1990); JUNK FAX LEGISLATION (1989); DIAL 900 LINES: TELEPHONES PAY TO PLAY (1988); LIVING ON THE LEADING EDGE: STATE POLICY ISSUES FOR EDUCATION AND ECONOMIC DEVELOPMENT IN A GLOBAL ECONOMY (1986); BREAKING UP IS HARD TO DO: THE STATES AND THE AT&T BREAKUP (1983); CELLULAR RADIO (1983); CABLE TELEVISION: STATE OR FEDERAL REGULATION (1982); STATE TELECOMMUNICATIONS ACTIVITIES (1980); STATE AND LOCAL FACTORS AFFECTING THE DEVELOPMENT OF PUBLIC SECTOR TELECOMMUNICATIONS APPLICATIONS (1979).

0848   **National Association of Regulatory Utility Commissioners** (NARUC). 1102 Interstate Commerce Commission Bldg., 12th St. and Constitution Ave. NW. Mail address: P.O. Box 684, Washington, DC 20044-0684. Voice: 202 898 2200. Fax: 202 898 2213. Nongovernmental organization of U.S. and Canadian federal, state, and municipal commissions that regulate utilities and carriers. Promotes state utilities commissions' regulatory uniformity, coordination, and cooperation to protect common public interest by study and discussion of issues and subjects concerning public utilities operation and supervision. Supports research of National Regulatory Research Institute (0859). Major offices include:
* General Counsel. Voice: 202 898 2200. Fax: 202 898 2213.
* Committee on Communications. Voice: 202 626 5110. Fax: 202 638 2330.
* Staff Subcommittee on Communications. Voice: 208 334 0316.

NARUC issues many publications on communication; telephone 202 898 2203 for an annotated listing. These include UNIFORM SYSTEM OF ACCOUNTS for cable, cellular, and radio services; AUDIT REPORTS on RBOC operations, DIRECTORY OF COMMUNICATIONS PROFESSIONALS, INTRASTATE TELECOMMUNICATIONS COMPETITION, MODEL TELECOMMUNICATIONS SERVICE RULES, REPORT ON THE REVIEW OF BELLCORE TECHNICAL WORK EFFORTS, SEPARATIONS MANUAL, and TELEPHONE SERVICE QUALITY HANDBOOK, among others.
NARUC regional organizations include:
1 **Great Lakes Conference of Public Utility Commissioners**. Voice: 301 333 6066.
2 **Mid-America Regulatory Commissioners**. Voice: 501 682 1451.
3 **New England Conference of Public Utility Commissioners, Inc**. Voice: 207 622 7694.
4 **Southeastern Association of Regulatory Utility Commissioners**. Voice: 615 341 3668.
5 **Western Conference of Public Service Commissioners**. Voice: 415 557 3474.

0849   **National Conference of State Legislatures**. 1560 Broadway, Ste. 700, Denver, CO 80202. Voice: 303 830 2200. Fax: 303 863 8003. Voices interests of states in federal decision-making process. Publishes studies and reports on regional uses and applications of telecommunications, including CABLE TELEVISION-COMPUTER PARTNERSHIP (1983) and TELECONFERENCING AND OTHER ELECTRONIC MEDIA USE IN THE LEGISLATURES (1980).

0850   **National Governor's Association**. Committee on Transportation Commerce and Communications. Hall of States, 444 N. Capital St. NW, Ste. 267, Washington, DC 20001. Voice: 202 624 5300. Fax: 202 624 5313. State governors' platform for addressing national policy. Also maintains Policy Research Center.

0851 **National League of Cities.** Transportation & Communications Committee. 1301 Pennsylvania Ave. NW, Washington, DC 20004. Voice: 202 626 3000. Fax: 202 626 3043. Interests in national municipal policy. Also maintains National Association of Telecommunications Officers and Advisors, voice: 202 626 3160, fax: 202 626 3103, consisting of 600 agencies and individuals in city and county government and regional authorities involved in development, regulation, and administration of cable television and other telecommunications services. Publishes newsletters and journals as well as studies and reference works, including TELECOMMUNICATIONS: THE NEXT AMERICAN REVOLUTION (1994), CRUISING THE INFORMATION SUPERHIGHWAY [video] (1994), and TELECOMMUNICATIONS SELF-ASSESSMENT: A MANUAL FOR LOCAL GOVERNMENT (1982).

6-D. Policy Research Centers and Organizations

See 0071.4. Advanced Research Projects Agency.

0852 **Catholic University.** Communications Law Institute. Law School, Washington, DC 20064. Voice: 202 319 5140. Publishes COMMLAW CONSPECTUS (1993-date, annual) and research studies, including EXPANDING THE ORBITAL ARC (1984) and NEW DEVELOPMENTS IN INTERNATIONAL TELECOMMUNICATIONS POLICY (1983), also sponsored by Federal Communications Bar Association (0574).

0853 **City University of New York.** Stanton Haskell Center for Public Policy & Telecommunications & Informations Systems. New York, NY 10036. Voice: 212 642 2984. Fax: 212 642 1959. Has published working papers as well as two major reports, BROADCAST OF TWO-WAY VIDEO IN DISTANCE EDUCATION (1993) and PROJECT T E L (TELECOMMUNICATIONS FOR EDUCATION AND LEARNING) (1994).

0854 **Columbia University.** CITI: Columbia Institute for Tele-Information. Graduate School of Business, New York, NY 10027. Voice: 212 854 4222. Fax: 212 932 7816. Well-established research program with many publications. Publishes annual report and a voluminous "Working Paper Series" (with over 600 titles to date), including C. Edwin Baker's MERGING PHONE AND CABLE (1994), Stanley M. Besen's RATE REGULATION, EFFECTIVE COMPETITION, AND THE CABLE ACT OF 1992 (1994), Michael Botein's U.S. TELECOMMUNICATIONS IN THE WORLD MARKET: CHANGING POLICY CONSIDERATIONS (1992), Eli M. Noam's THINKING ABOUT A WIRELESS FUTURE: THE SOCIAL IMPLICATIONS OF MOBILE COMMUNICATIONS (1994), and Donald W. Hawthorne's REWIRING THE FIRST AMENDMENT: MEANING, CONTENT, AND PUBLIC BROADCASTING (1994), and many others.

See 0760.1. **Congressional Research Service.**

0855 **Harvard University.** Harvard Program on Information Resources Policy. 200 Aiken Hall, Cambridge, MA 02138. Voice: 617 495 4114. Internet: PIRP@DAS.HARVARD.EDU. Publishes research studies on telecommunications and information, including Patricia Hirl Longstaff's INFORMATION THEORY AS A BASIS FOR RATIONAL REGULATION OF THE COMMUNICATIONS INDUSTRY (1994), CELLULAR TELEPHONES: IS THERE REALLY COMPETITION (1994), Tommy T. Osborne's BETTER TELEPHONE SERVICE FOR THE HAVE NOTS: IN WHOSE INTEREST, BY WHICH MEANS, AND WHO PAYS? (1992); Robert A. Travis' THE TELECOMMUNICATIONS INDUSTRY IN THE U.S. AND INTERNATIONAL COMPETITION: POLICY VS. PRACTICE (1990); Naoyuki Koike's CABLE TELEVISION AND TELEPHONE COMPANIES: TOWARD RESIDENTIAL BROADBAND COMMUNICATIONS SERVICES IN THE UNITED STATES AND JAPAN (1990); Satoshi Shinoda's COMPETITION IN LOCAL TELECOMMUNICATIONS MARKETS: THE U.S. AND JAPAN (1989); Karin Leonard Sonneman's LOUISIANA VS. FCC: ITS IMPLICATIONS FOR THE BALANCE OF POWER BETWEEN STATE AND FEDERAL AUTHORITIES IN THELECOMMUNICATIONS REGULATION (1989); Carol L. Weinhaus' TELECOMMUNICATIONS INDUSTRY: TACTICAL

DISPUTES, ELEMENTS OF CHANGE, AND STRATEGIC OUTCOMES (1989); Vincent Marier's U.S. COMMUNICATIONS POLICY: A SURVEY AND DATABASE OF EXECUTIVE ORDERS AND CONGRESSIONAL ACTS OVER 24 YEARS (1986); and many others.

0856   **Massachusetts Institute of Technology.** Research Program on Communications Policy. Cambridge, MA 02139. Voice: 617 253 4138. Fax: 617 253 7326. Internet: RPCP@FARN-SWORTH.MIT.EDU. Also World-Wide Web: gopher://farnsworth.mit.edu//11/rpcp. Well established telecommunications and information technology research program in MIT's Center for Technology, Policy, and Industrial Development. Focuses on technical, economic, and policy issues. Organizes "Cambridge Roundtable" discussion forum, hosts and participates in workshops, seminars, and other programs. Publishes many studies and reports in "Seminar Notes" series, including Ithiel de Sola Pool's CROSS OWNERSHIP POLICY IN A CHANGING MEDIA ENVIRONMENT (1993), Lisa Allen Vawter's DEREGULATION OF CABLE (1984), Albert Halprin's TELEPHONE ACCESS CHARGES (1983), and Charles L. Jackson's THE CHANGING BROADCAST SPECTRUM: SCA, LPTV, DBS, & MDS (1983).

0857   _____. Research Program in Telecommunications. Sloan School of Management, Cambridge, MA 02139. Voice: 617 253 3644.

0858   **Michigan State University.** Institute of Public Utilities. Graduate School of Business Administration, 113 Olds Hall, East Lansing MI 48824-1047. Voice: 517 355 1876. Promotes teaching and research on public utilities, particularly regulation. Courses, seminars, conferences, publications. Members include 38 private sector utilities companies. Publishes PROCEEDINGS. Two annual conferences—one in East Lansing and a December meeting in Williamsburg for state PUC personnel.

0859   **National Regulatory Research Institute.** (NRRI) Columbus, OH 43210-1002. Voice: 614 292 9404. Fax: 614 292 7196. Research center of the National Association of Regulatory Utility Commissioners (0848). Publishes NATIONAL REGULATORY RESEARCH INSTITUTE QUARTERLY BULLETIN (1980–date), which abstracts and indexes selected decisions of state regulatory commissions. Also publishes proceedings of biennial NARUC conferences. Conducts, sponsors, and publishes research studies on telecommunications economics and regulatory issues, including COMPETITION AND INTERCONNECTION: THE CASE OF PERSONAL COMMUNI-CATIONS SERVICES (1994); John D. Borrows' UNIVERSAL SERVICE IN THE UNITED STATES: DIMENSIONS OF DEBATE (1994); Phyllis Bernt's THE IMPACT OF ALTERNATIVE TECHNOL-OGIES ON UNIVERSAL SERVICE AND COMPETITION IN THE LOCAL LOOP (1992); Robert J. Graniere's INTRASTATE BASIC SERVICE ELEMENTS: EFFECTS ON THE PRICES OF MESSAGE TOLL SERVICE AND PLAIN OLD TELEPHONE SERVICE (1991); Patricia D. Kravtin's A PUBLIC GOOD/PRIVATE GOOD FRAMEWORK FOR IDENTIFYING POTS OBJECTIVES FOR THE PUBLIC SWITCHED NETWORK (1991); William G. Shepherd's DOMINANCE, NON-DOMINANCE, AND CONTESTABILITY IN A TELECOMMUNICATIONS MARKET: A CRITICAL ASSESSMENT (1990); Raymond Lawton's FACTORS AFFECTING THE DEFINITION OF THE LOCAL CALLING AREA: AN ASSESSMENT OF TRENDS (1990); and many more.

0860   **National Research Council.** 2101 Constitution Ave., Washington DC 20418. Voice: 202 334 2000. Private organization chartered, but not supported, by Congress. Serves as independent advisor to government on scientific and technical issues. Membership includes representatives of telecommuni-cations industry, academic, and research communities.

    1 **Publications and Administrative Services.** Voice: 703 306 2003

    2 **Office of Science & Technology Intrastructure.** Voice: 703 306 1040

    3 **Directorate for Computer & Information Science & Engineering.** Voice: 703 306 1900.

    4 **Division of Information, Robotics & Intelligent Systems** Voice: 703 306 1930. Specifically: Interactive Systems (Voice: 703 306 1928), and Information Technology & Organizations (Voice: 703 306 1927)

    5 Division of Networking & Communications Research & Infrastructure, Voice: 703 306 1950

    6 Division of Electrical and Communications Systems, Voice: 703 306 1390

0861   **National Science Foundation.** 4201 Wilson Blvd., Arlington, VA 22230, voice: 703 306 1234; fax: 703 306 0202. Created by President Truman in 1950, NSF is the federal government's primary means of funding scientific and engineering research. In 1994 it moved out of Washington into the suburbs.

    1 Office of Science and Technology Infrastucture. Voice: 703 306 1040

    2 Office of Computer and Information Science and Engineering. Voice: 703 306 1900

    3 National Science Board. Voice: 703 306 2000.

0862   **New Mexico State University.** Center for Public Utilities. College of Business Administration and Economics, Box 3001, Department 3MPD, Las Cruces NM 88003. Voice: 505 646 3242. Fax: 505 646 6025. Academic education and training programs in telecommunications regulation for utilities industry and government. Sponsors seminars and conferences. No publication program.

0863   **New York Law School.** Communications Media Center, 57 Worth St., New York, NY 10013. Voice: 212 431 2160. Fax: 212 966 2053. Sponsors lectures, symposia, conferences, and academic programs, including Media Law Project, which publishes MEDIA LAW AND POLICY (1991-date, semiannual), formerly MEDIA LAW JOURNAL. Also co-sponsored annual program with Electronics International Corporation, the European Union's DG XIII, and the Japanese Ministry of Post and Telecommunications and co-published proceedings, COOPERATION AND COMPETITION IN TELECOMMUNICATIONS (1993-date).

0864   **Northwestern University.** Annenberg Washington Program on Communication Policy Studies. 1455 Pennsylvania Ave, NW, Washington, DC 20004. Voice: 202 393 7100. Fax: 202 638 2745. Supports programs focusing on full range of topics, from emergency communications systems and organ donor notification networks to National Information Infrastructure and international regulations. Conferences and workshops plus related publications, including Dale Hatfield's SPEEDING TELEPHONE SERVICE TO RURAL AREAS (1994) and Margaret Gerteis' VIOLENCE, PUBLIC HEALTH, AND THE MEDIA (1994). Also publishes ANNUAL REPORT (1990–date), continuing ANNENBERG WASHINGTON PROGRAM: A REPORT ON THE FIRST SIX YEARS, 1983–1989 (1989).

See 0766. **Office of Science and Technology Policy.**

See 0761. **Office of Technology Assessment.**

0865   **Ohio State University.** Center for Advanced Study in Telecommunications. Columbus, OH 43210-1286. Voice: 614 292 8444. Fax: 614 292 2055. Internet: cast@eng.ohio-state.edu. Sponsors research and programs focusing on telemedicine, distance education, and international telecommunications policy, as well as national policy issues, like privacy, rates, and access. Sponsors annual distance education symposia and other conferences. Publishes CAST CALENDAR AND NEWSLETTER (1989–date, bimonthly), conference proceedings, and "Working Files," including Dirk Scheerhorn's COMPUTER-BASED TELECOMMUNICATIONS AMONG AN ILLNESS-RELATED COMMUNITY (1994), Stephen Acker's USING DISTANCE EDUCATION TO SPAN CULTURES (1994), and Scott Patterson's RURAL USERS' EXPECTATIONS OF THE INFORMATION SUPERHIGHWAY (1994), among others.

0866   **University of Florida.**

    1 Brechner Center for Freedom of Information. Journalism Bldg., Gainesville, FL 32611. Voice: 904 392 2273. Fax: 904 392 3919. Research on telecommunications and media policy. Publishes BRECHNER REPORT (1977-date).

    2 Public Utility Research Center. 224 Matherly Hall, Gainesville, FL 32611. Voice: 904 392 0151. Fax: 904 397 6250. Supported by Florida Public Services Commission. Research on telecommunications policy and other utilities; participates in Telecommunications Industry Analysis

Project. Publishes PURC REVIEW (1971-date) as well as working papers, including a 20-year report on center.

0867   **University of Hawaii**. East-West Center. Honolulu HI 96822. Voice: 808 944 7111. Supports extensive publication program focusing on telecommunications and larger development issues in Asia and Pacific. Publications sponsored by East-West Center include INFORMATION TECHNOLOGY AND GLOBAL INTERDEPENDENCE (New York: Greenwood, 1989); THE SOCIAL AND ECONOMIC ISSUES OF TELEPHONES: AN ANNOTATED BIBLIOGRAPHY AND SUMMARY PAPER (1984).

6-E. Selected Secondary Resources:

6-E-1. Surveys

0868   Bolter, Walter G., James W. McConnaughey, and Fred J. Kelsey. **TELECOMMUNICA-TIONS POLICY FOR THE 1990s AND BEYOND**. Armonk, NY: M.E. Sharpe, 1990, 435 pp. Useful economic analysis of (primarily FCC) deregulation of the 1980s and likely policy directions for the 1990s.

0869   Brock, Gerald, W. **TELECOMMUNICATION POLICY FOR THE INFORMATION AGE**. Cambridge, MA: Harvard University Press, 1994, 324 pp. Former FCC official reviews major Commission proceedings of the 1980s and early 1990s with useful analysis.

0870   Brotman, Stuart N. **THE TELECOMMUNICATIONS DEREGULATION SOURCEBOOK**. Norwood, MA: Artech House, 1987, 342 pp. Brief excerpts with notes on key decisions with sections on broadcasting, cable and common carrier.

0871   Crandall, Robert W. **AFTER THE BREAK-UP: U.S. TELECOMMUNICATIONS IN A MORE COMPETITIVE ERA**. Washington, DC: Brookings, 1991, 174 pp. Brief six-chapter survey, including service and rates, equipment market, and income distribution and economic welfare.

0872   Danielsen, Albert L., and David R. Kamerschen. **TELECOMMUNICATIONS IN THE POST-DIVESTITURE ERA**. Lexington, MA: Lexington Books, 1986, 252 pp. Twenty-one analyses comparing and contrasting the viewpoints of regulators, companies, the pricing of telecommunications services, and related concerns. Notes, index.

0873   Elton, Martin C.J., ed. **INTEGRATED BROADBAND NETWORKS: THE PUBLIC POLICY ISSUES**. Amsterdam: North-Holland, 1991. Sixteen papers based on a two-year research project cover technical, economic and more general policy aspects.

0874   Sapronov, Walter, ed. **TELECOMMUNICATIONS AND THE LAW**. Rockville, MD: Computer Sciences Press, 1988, 443 pp. Collection of 20 articles on different aspects of regulation, some fairly dated even on publication. Notes.

0875   Tunstall, Jeremy. **COMMUNICATIONS DERGULATION: THE UNLEASING OF AMERICA'S COMMUNICATIONS INDUSTRY**. Oxford, England: Basil Blackwell, 1986, 324 pp. A British expert views the dramatic changes in American media and telecommunications, carefully assessing the roles of different regulatory players. Notes, bibliography, index.

6-E-2. Treatises and Texts

0876   Brenner, Daniel L. **LAW AND REGULATION OF COMMON CARRIERS IN THE COMMUNICATION INDUSTRY**. Boulder, CO: Westview, 1992, 299 pp. Basic casebook treatment with commentary with sections on public utility theory, Title II regulation, federal and state jurisdiction, price caps, dominant and nondominant carriers, RBOC restrictions, access and bypass, deregulated markets and the like.

0877   Kellogg, Michael K., John Thorne, Peter W. Huber. **FEDERAL TELECOMMUNICA-TIONS LAW**. Boston: Little, Brown, 1992, 914 pp. **1993 SUPPLEMENT**. 1993, 149 pp. Detailed coverage of a fast-changing field in 16 chapters including economic regulation, international issues, the relationship of cable and telephone industries, and privacy.

0878   Kennedy, Charles H. **AN INTRODUCTION TO U.S. TELECOMMUNICATIONS LAW.** Norwood, MA: Artech House, 1994, 169 pp. One of the most concise discussions available, this offers nine chapters on local exchange carriers (half the book), and "other players" including interexchange and specialized players. Notes, tables, index.

0879   Newberg, Paula R., ed. **NEW DIRECTIONS IN TELECOMMUNICATIONS POLICY.** Durham, NC: Duke University Press, 1989 (two vols), 414 + 346 pp. Very useful set of original papers assessing regulatory policy in both media and telephony (vol. 1), and information and economic policies (vol. 2). Notes, index.

6-E-3. Access, Privacy and Intellectual Property

0880   Braverman, Burt A., and Frances J. Chetwood. **INFORMATION LAW: FREEDOM OF INFORMATION PRIVACY, OPEN MEETINGS AND OTHER ACCESS LAWS.** New York: Practising Law Institute, 1985, two loose-leaf vols. Highly detailed (designed as advice for attorneys) information on these laws. Notes.

0881   Denning, Dorothy E., and Herbert S. Lin, eds. **RIGHTS AND RESPONSIBILITIES OF PARTICIPANTS IN NETWORKED COMMUNITIES.** Washington: National Academy Press, 1994, 160 pp. Report drawn from 1992–1993 workshops and fora sponsored by the National Research Council's Computer Science and Telecommunications Board and other committees. Examines policies, laws, regulations, and ethical standards of users and providers of computer-mediated and other electronic services, focusing especially on issues of free speech, intellectual property, privacy, and computer-related crime.

0882   Office of Technology Assessment (U.S. Congress). **INTELLECTUAL PROPERTY RIGHTS IN AN AGE OF ELECTRONICS AND INFORMATION.** Washington, DC: GPO, 1986, 299 pp. Detailed review of policy options in keeping copyright and patents current amidst changing technology. Tables, charts, notes.

0883   Privacy Protection Study Commission. **PERSONAL PRIVACY IN AN INFORMATION SOCIETY.** Washington: GPO, 1977, 654 pp. The final report of the Presidential group which studied the question for two years. Among subject areas explored are mailing lists, consumers and credit, insurance, employment records, medical records, and government record-keeping. The appendices to the report were published separately:

    1 **PRIVACY LAW IN THE STATES,** 85 pp.
    2 **THE CITIZEN AS TAXPAYER,** 57 pp.
    3 **EMPLOYMENT RECORDS,** 100 pp.
    4 **THE PRIVACY ACT OF 1974: AN ASSESSMENT,** 173 pp.
    5 **TECHNOLOGY AND PRIVACY,** 88 pp.

0884   Rubin, Michael Rogers. **PRIVATE RIGHTS, PUBLIC WRONGS: THE COMPUTER AND PERSONAL PRIVACY.** Norwood, NJ: Ablex, 1988, 153 pp. Evolution of privacy safeguards, forces of change, the individual under assult, regulation of computer records in the U.S., and changes needed. References, index.

0885   Saxby, Stephen, ed. **ENCYCLOPEDIA OF INFORMATION TECHNOLOGY LAW.** London: Sweet & Maxwell, 1990, two loose-leaf vols. Focusing primarily on British practice, this includes such topics as electronic funds transfer, computer contracts, liability, risk management, employment and taxation aspects. Notes, index.

0886   Wright, Benjamin. **THE LAW OF ELECTRONIC COMMERCE: EDI, FAX, AND E-MAIL: TECHNOLOGY, PROOF, AND LIABILITY.** Boston: Little, Brown, 1991, 432 pp. A pioneering analysis of the relationship of often-old information policy law with fast-changing electronic means of communicating.

6-E-4. Convergence of Media/Telecommunication

0887   Branscomb, Anne Wells. **WHO OWNS INFORMATION? FROM PRIVACY TO PUBLIC ACCESS**. New York: Basic Books, 1994, 241 pp. A noted researcher and attorney assesses the stresses and strains caused by changing technology on often old legal protections.

0888   Brotman, Stuart N. **TELEPHONE COMPANY AND CABLE TELEVISION COMPE-TITION: KEY TECHNICAL, ECONOMIC, LEGAL, AND POLICY ISSUES**. Norwood, MA: Artech House, 1990, 509 pp. Anthology of some 25 articles, primarily from the 1980s, that shed light on the relationship.

0889   de Sonne, Marcia, ed. **CONVERGENCE: TRANSITION TO THE ELECTRONIC SUPERHIGHWAY**. Washington: National Association of Broadcasters, 1994, 260 pp. Chapters on shaping the elctronic video future, enabling technologies, getting to the 500-channel universe, opportunity and peril on the new electronic frontier, financial implications of telco/video wars, the collision of industries, concerns about audience habits--and willingness to pay, and multimedia broadcasting. Charts, tables, notes.

0890   Elton, Martin, C.J., ed. **INTEGRATED BROADBAND NETWORKS: THE PUBLIC POLICY ISSUES**. Amsterdman: North-Holland, 1991, 360 pp. Sixteen papers assess new services, their technological and economic viability, and institutional and regulatory issues. Glossary, index.

0891   Gasman, Lawrence. **TELECOMPETITION: THE FREE MARKET ROAD TO THE INFORMATION HIGHWAY**. Washington: The CATO Institute, 1994, 177 pp. A libertarian argument proposing a model of the process and chapters on the myth of the communications monopoly, who should own the airwaves, the information infrastructure, what is to be done, and whither the FCC. Index, notes.

0892   Johnson, Leland. **TELEPHONE COMPANY ENTRY INTO CABLE TELEVISION: COMPETITION, REGULATION, AND PUBLIC POLICY**. Santa Monica, CA: Rand Corp., 1992, 60 pp. See next entry.

0893   _____, and David P. Reed. **RESIDENTIAL BROADBAND SERVICES BY TELEPHONE COMPANIES? TECHNOLOGY, ECONOMICS AND PUBLIC POLICY**. Santa Monica, CA: Rand Corp., 1990, 104 pp. Major results of Rand studies, these two reports detail the case for and against telephone provision of information/media services.

0894   **TELECOMMUNICATIONS AND BROADCASTING: CONVERGENCE OR COLLISION?** Paris: Organization of Economic Cooperation and Development, 1992, 287 pp. Technical trends, service convergence, corporate strategies, service regulation and deregulation, and details on trends in Europe. Figures, bibliography, glossary.

6-E-5. National Information Infrastructure

0895   Information Infrastructure Task Force. **THE NATIONAL INFORMATION INFRA-STRUCTURE: AGENDA FOR ACTION**. Washington: National Telecommunications and Information Administration, 1993, 26 pp. Initial Clinton Administration statement on the NII initiative with plans for the all-government task force and major research thrusts.

0896   _____. **NATIONAL INFORMATION INFRASTRUCTURE: PROGRESS REPORT, SEPTEMBER 1993-1994**. Washington: GPO, 1994, ca 100 pp. Details the Clinton Administration's first year of research, policy shaping, and legislative attempts to facilitate NII development and operation.

0897   Information Infrastructure Task Force Committee on Applications and Technology. **PUTTING THE INFORMATION INFRASTRUCTURE TO WORK**. Washington: National Institute of Standards and Technology (SP 857), 1994, 109 pp. Initial substantive report attempting to set NII agenda to ensure widespread use and application of changing technologies.

0898   **20/20 VISION: THE DEVELOPMENT OF A NATIONAL INFORMATION INFRA-STRUCTURE.** Washington: National Telecommunications and Information Administration (SP 94-28)/NTIS, 1994, 163 pp. Collection of 15 papers by as many experts, all exploring aspects of how to develop the NII and what it can do once it is in place. Notes, charts.

6-E-6. State and Rural Area Policies

0899   Baumol, William J. and J. Gregory Sidak. **TOWARD COMPETITION IN LOCAL TELEPHONY.** Cambridge, MA: MIT Press, 1994, 169 pp. An economic study of how to improve state and local regulation of telephone competition, including policies to promote entry of new carriers.

0900   Cohen, Jeffrey E. **THE POLITICS OF TELECOMMUNICATIONS REGULATION: THE STATES AND THE DIVESTITURE OF AT&T.** Armonk, NY: M.E. Sharpe, 1992, 177 pp. Historical treatment of the changing regulatory role of states and their public utility commissions before and since the divestiture, showing how state roles have increased in importance.

0901   Horning, John S., et al. **EVALUATING COMPETITIVENESS OF TELECOMMUNI-CATIONS MARKETS: A GUIDE FOR REGULATORS.** Columbus, OH: National Regulatory Research Institute, 1988. This research arm of NARUC issues several t'com-related titles every year—this one is of broad general value.

0902   Johnson, Leland L., and David P. Reed. **RESIDENTIAL BROADBAND SERVICES BY TELEPHONE COMPANIES? TECHNOLOGY, ECONOMICS, AND PUBLIC POLICY.** Santa Monica, CA: Rand, 1990, 104 pp. Assesses new residential construction, other network architectures, competition and government regulation, service to existing residential areas, and public policy implications. Tables, appendices, notes.

0903   Mueller, Milton L. **TELEPHONE COMPANIES IN PARADISE: A CASE STUDY IN TELECOMMÜNICATIONS DEREGULATION.** New Brunswick, NJ: Transaction Publishers, 1993, 185 pp. Assesses the almost total deregulation of telephone company practices in the state of Nebraska and determines the resulting lack of government policy has had little overall impact on telephone company operations.

0904   Parker, Edwin, and Heather Hudson, et al. **ELECTRONIC BYWAYS: STATE POLICIES FOR RURAL DEVELOPMENT THROUGH TELECOMMUNICATIONS.** Boulder, CO: Westview, 1992, 306 pp.

0905   Schmandt, Jurgen, et al. **TELECOMMUNICATIONS AND RURAL DEVELOPMENT: A STUDY OF PRIVATE AND PUBLIC SECTOR INNOVATION.** New York: Praeger, 1991, 272 pp.

0906   Teske, Paul Eric. **AFTER DIVESTITURE: THE POLITICAL ECONOMY OF STATE TELECOMMUNICATIONS REGULATION.** Albany: State University of New York Press, 1990, 162 pp.

6-D-7. City Policies

0907   Dutton, William H., et al., eds. **WIRED CITIES: SHAPING THE FUTURE OF COMMUNICATIONS.** Boston: G.K. Hall, 1987, 492 pp. Twenty-eight original papers on developments in the U.S. and elsewhere. A very useful comparative survey.

0908   Schmandt, Jurgen, et al., eds. **THE NEW URBAN INFRA-STRUCTURE: CITIES AND TELECOMMUNICATIONS.** New York: Praeger, 1990, 344 pp. Detailed case studies of 12 American and Canadian cities and their application of telecommunication to city governmental, industrial, and educational needs.

International

Chapter 7 differs from the rest of the book in that it largely ignores the domestic telecommunications industry. Instead, it includes resources and international and foreign national organizations that help define the role of the United States in international telecommunications. As noted above, we identify many indexes, abstracts, and other resources that analyze U.S. telecommunications from international and foreign national perspectives. The International Telecommunication Union (ITU) (0950), the world's major international telecommunications organization, is also detailed here—including information about its 1993 restructuring, which in turn is reflected in U.S. agenices and offices. Additionally, Section 7-D includes resources and organizations in international and foreign national satellite communications. This chapter concludes with statistical resources on international telecommunications management and the U.S. role overseas.

7-A. Bibliographic Resources

7-A-1. Bibliographies

0909   Asian Mass Communication Research and Information Centre. **COMMUNICATION POLICIES AND PLANNING IN ASIA: AN ANNOTATED BIBLIOGRAPHY**. Singapore: AMIC, 1978, 131 pp. Some 354 selected annotated entries. Author, title, country index.

0910   **BIBLIOGRAPHY OF PUBLICATIONS IN COMMUNICATION SATELLITE TECHNOLOGY AND POLICY**. Washington: INTELSAT, May 1994, 13 pp. Semi-annotated listing drawn from INTELSAT library holdings. Includes reference sources, journals, directories and dictionaries, and books divided into subject categories. Some items are annotated. (Regularly updated.)

0911   **BRIDGING THE GAP III: A GUIDE TO TELECOMMUNICATIONS AND DEVELOPMENT**. Washington: INTELSAT, 1987, 118 pp. Classified annotated listing covering socioeconomic development, rural and agricultural development, health, tele-education, training and assistance, communication technology, and global applications (arranged by geographic regions). Emphasis on developing nations.

0912   **COMMUNICATION POLICIES AND PLANNING: AN ANNOTATED BIBLIOGRAPHY**. Budapest: Mass Communication Research Centre, 1986, 79 pp. "Prepared by the International Association of Mass Communication Research (IAMCR) for the United Nations Educational, Scientific, and Cultural Organization (UNESCO)." Classified annotated listing: relevant items under "International Policy Studies," "Regional Studies," "Country Studies," "Telecommunication," "Computer/data," and "Space." A useful and important compilation.

0913   Fackelman, Mary P. and Kimberly A. Krekel. **INTERNATIONAL TELECOMMUNICATIONS BIBLIOGRAPHY**. Washington: GPO (Department of Commerce, Office of Telecommunication Special Report 76–7), 1976, 176 pp. Annotations are often quite extensive in this subject-divided resource. Author and subject indexes.

0914   **INFORMATION TECHNOLOGY AND THE LAW: AN INTERNATIONAL BIBLIOGRAPHY**. A H Dordrecht, Netherlands: Martinus Nijhoff, 1992–date, biannual. Earlier titles include INFORMATICA E DIRITTO and BIBLIOGRAFIA INTERNAZIONALE D'INFOR-MATICA E DIRITTO (1975–1991). "Edited by the Instituto per la Documentazione Giuridica of the Consiglio Nazionale delle Richerche, Florence, Italy." Major comprehensive bibliography for computer and information technology applications in law, regulation, legislation, and policy, as well as for legal aspects of uses of computers and information technology in society (business, industry, education, government, etc.). Annotated entries for about 1,000 items per issue (books, court decisions and judgments, legislation, journal articles, databases) in classified subject arrangement: class 0.4 and 8.2 cover reference works, 5.2 "Computers in industry", 5.2.2. "Computers in communications industries," for example. Author and keyword indexes: extensive listings under names of companies and "Tele-" words. Valuable coverage of major English and non-English language telecommunication law, policy, and research journals, including COMMUNICATIONS AND THE LAW (1109), TELECOM-MUNICATIONS POLICY (1123), French, Russian, German, Dutch, Spanish, Italian, and Scandinavian journals.

0915   Leteinturier, Christine. **COMMUNICATION ET MEDIAS: GUIDE DES SOURCES DOCUMENTAIRES FRANCAISES ET INTERNATIONALES**. Paris: Editions Eyrolles, 1991, 158 pp. Included because it has no real English-language equivalent, except perhaps WORLD COMMUNICATION REPORT (0929), for international communication resources. Guide to French and international communications information centers and organizations, bibliographies, studies, journals, statistical sources, dictionaries, academic programs, etc. General index. Merits replication worldwide.

0916   Melody, William H., Dallas W. Smythe, Robin E. Mansell, and Ursel Koebberling. **COMMUNICATION, INFORMATION, AND CULTURE: ANNOTATED BIBLIOGRAPHY**. Vancouver, BC: Department of Communication, Simor Fraser University, and Transnational Research Institute for Policy (TRIP), in conjunction with Centro de Estudios Sobre Cultura Transnacional, Instituto Para America Latina (IPAL), Lima, Peru, 1985, 213 pp. Classified, annotated entries for books and articles from 1960–1985. Listings for telecommunications in chapter for law and regulation, political and economic dependency, communication and development, information society, etc. Author index.

0917   **OUTER SPACE: A SELECTIVE BIBLIOGRAPHY; L'ESPACE EXTRA-ATMOSPHERIQUE: BIBLIOGRAPHIE SELECTIVE**. New York: United Nations, 1981, 123 pp. "Prepared by Dag Hammarskjold Library." Extensive listings on satellite technology and communications in 3 sections covering books and articles, 1971–1981; UN documents, including ITU, Unesco, WIPO publications; and publications in Russian for 1971–1981.

0918   Pisarek, Walery. **ANNOTATED BIBLIOGRAPHY ON COMMUNICATION POLICIES IN BULGARIA, CZECHOSLOVAKIA, GDR, HUNGARY, POLAND, RUMANIA, USSR, AND YUGOSLAVIA**. Cracow, Poland: Central European Mass Communication Research Centre, 1985, unpaged. Describes legal and official documents and scholarly books and articles in classified subject arrangement under each country.

0919   _____. **INTERNATIONAL BIBLIOGRAPHY OF MASS COMMUNICATION BIBLIOGRAPHIES**. Cracow, Poland: Bibliographic Section of the International Association of Mass Communication Research/Press Research Centre R.S.W. "Prasa" in Cracow, 1972, unpaged. "Draft version." Perhaps the only comprehensive (though certainly dated) listing of bibliographies in mass communications. Arranged by country.

0920   Rahim, Syed. **COMMUNICATION POLICY AND PLANNING FOR DEVEL-OPMENT: A SELECTED ANNOTATED BIBLIOGRAPHY**. Honolulu: East-West Communication Institute, 1976, 285 pp. Some 395 well-annotated entries on aspects of communication policy and planning in 53 countries. Author, country, subject indexes: relevant listings under "Computer," "Satellite," "Telecommunications," others.

0921   Richstad, Jim, and Jackie Bowen. **INTERNATIONAL COMMUNICATION POLICY AND FLOW: A SELECTED ANNOTATED BIBLIOGRAPHY**. Honolulu: East-West Communication Institute, 1976, 103 pp. Some 223 annotated entries. Author and subject indexes: useful headings include "Satellite communication," "Telecommunications systems."

0922   Servaes, Jan, and Mario Simons. **COMMUNICATION POLICIES IN EUROPE: A SELECTED AND ANNOTATED BIBLIOGRAPHY.** Leuven: Centrum voor Communicatiewetensschappen, Katholicke Universiteit Leuven; Mass Communication Research Centre, Budapest, 1987, 204 pp. Classified annotated (in both French and English) listing of 491 items. Covers "works of reference, official texts, policy documents and reports." "International" chapter includes many items related to U.S. Author and subject indexes: relevant listings under "Telecommunication," "Telematics," and elsewhere.

0923   Snow, Marcellus S., and Meheroo Jussawalla. **TELECOMMUNICATION ECONOMICS AND INTERNATIONAL REGULATORY POLICY: AN ANNOTATED BIBLIOGRAPHY.** Westport, CT: Greenwood, 1986, 216 pp. About 500 well-annotated entries in three-part arrangement with topical subchapters; see especially section 12, "Studies of Telecommunications Regulation in the United States." Subject and author indexes.

0924   Talbot, Dawn E. **JAPAN'S HIGH TECHNOLOGY: AN ANNOTATED GUIDE TO ENGLISH-LANGUAGE INFORMATION SOURCES.** Phoenix, AZ: Oryx, 1991, 171 pp. Describes abstracts and indexes, bibliographies, dictionaries, directories and handbooks, yearbooks, journals, industry reports, newspapers, databases, patents, and translations in telecommunications, communications, and electronics fields. Author, title, publisher, subject indexes.

0925   **TRANSNATIONAL CORPORATIONS: A SELECTIVE BIBLIOGRAPHY; LES SOCIETES TRANSNATIONALES: BIBLIOGRAPHIE SELECTIVE, 1991–1992.** New York: United Nations, 1993, 727 pp. See entry 0927 below.

0926   **TRANSNATIONAL CORPORATIONS: A SELECTIVE BIBLIOGRAPHY, 1988– 1990.** New York: United Nations, 1991, 617 pp. See next entry.

0927   **TRANSNATIONAL CORPORATIONS: A SELECTIVE BIBLIOGRAPHY, 1983– 1987.** New York: United Nations, 1987, 2 vols. Sponsored by UNCTAD Programme on Transnational Corporations; sponsored by UN Centre on Transnational Corporations (1975–1992) and Transnational Corporations and Management Division of the UN Department of Economic and Social development (1992–1993). Describes books and articles in classified arrangement with chapters on foreign direct investment; transnational corporations, their management and impacts on home and host countries; economic, political, social, international and national legal and policy issues of TNCs; contractual agreements of TNCs and governments. Author and subject indexes: relevant listings under "Cable television," "Information technology," "Satellite communications," and "Tele-" words.

0928   Wallace, Cynthia Day. **FOREIGN DIRECT INVESTMENT AND THE MULTINA- TIONAL ENTERPRISE: A BIBLIOGRAPHY.** Dordrecht, Netherlands: Martinus Nijhoff, 1988, 355 pp. Chapter 21, "Transborder Data Flows," includes 15 items (pp. 295–96); chapter 23 (pp. 301– 55) covers the U.S. and lists many items on the information and telecommunications industries. Unannotated; no subject index.

0929   **WORLD COMMUNICATION REPORT.** Paris: Unesco, 1989, 551 pp. Extensive surveys of communication and national and international economic development cooperation and assistance, recent developments in communications technologies, industrial growth patterns in communications, employment and training, information flow, legal and regulation and policy developments, and applications, with bibliographies of reference resources, directories of organizations and agencies, works, and statistics. Information relevant to international telecommunication throughout. Index; glossary of acronyms.

7-A-2. Selected Abstracting, Indexing, and Electronic Database Services

0930   **COMMUNICATION CONTENTS SISALLOT.** Tampere, Finland: University of Tampere and Nordicom, 1985–date, biannual. Reproduces tables of contents of more than 50 important communication journals from U.S., United Kingdom, France, Germany, Russia, Sweden, Norway, Denmark, Spain, Kenya, Singapore, Australia. A useful service for access to otherwise unindexed international journals.

0931   **EXCERPTA INFORMATICA: AN ABSTRACT JOURNAL OF RECENT LITER-ATURE ON AUTOMATION.** Tilburg, The Netherlands: Tilburg University Press and Tilburg University Library, 1985–date, monthly. Online version available on Datanet-1, Surfnet, and via Internet, with document order system directly from Tilburg University Library: internet docdel@kub.nl. Classified indexing for about 400 major English and non-English language policy and research journals. Covers about 50,000 items to date. English annotations. Author and keyword indexes: relevant listings under "tele-" words.

0932   **INDICATOR CONTENTS.** Calgary, Alberta: International Federation of Communication Associations, 1992–date, 3/year. First issues reprinted tables of contents of nearly 100 major international communication research journals, with the intention of including at least another 100 in subsequent issues. Titles from Australia, Brazil, Canada, Croatia, Ecuador, Germany, Italy, Japan, Netherlands, Singapore, Sweden, U.K., and U.S. A unique service.

0933   **INTERNATIONAL POLITICAL SCIENCE ABSTRACTS.** Oxford: Blackwell, 1951–date, quarterly. Comprehensive coverage of international political science literature in classified arrangement: relevant listings under "Information," "Mass media," "Television," and others.

0934   **MASS COM PERIODICAL LITERATURE INDEX.** Singapore: Documentation Unit, Asian Mass Communication Research & Information Centre, 1982–date, semiannual. Continues AMIC INDEX TO PERIODICALS (1972–1980) and MASS COM PERIODICAL LITERATURE (1981). Classified listing of articles featured in Asian trade newsletters and major research journals (many covered nowhere else), including ASIAN COMMUNICATIONS, TELECOM ASIA, NEW BREEZE, MOBILE ASIA PACIFIC, ASIA PACIFIC TELECOMMUNICATIONS, NEW ERA OF TELECOMMUNICATIONS IN JAPAN. Relevant listings include "Answering Machines," "Cable," "Cellular Communication," "Digital Broadcasting," "HDTV," "Satellites," "Telecommunications," "Teleconferencing," "Telephone," "Teleshopping," "Telework," VSAT." Coverage overlaps with AMIC's AMICNET online database, available by subscription, with document delivery services.

0935   **NORDICOM: BIBLIOGRAPHY OF NORDIC MASS COMMUNICATION LITER-ATURE; BIBLIOGRAFI OVER NORDISK MASSENKOMMUNIKATIONS-LITTERATUR.** Aarhus, Denmark: Nordic Documentation Center for Mass Communication Research, 1975–date, annual. Online version available from University of Aarhus via Internet. Comprehensive bibliography of mass communication research literature published in Denmark, Finland, Iceland, Norway, and Sweden. Vol. 1, "Document List," includes numbered bibliographic citations. Non-English language titles translated. Abstracts in language of publication. Vol. 2, "Index," includes English language keywords/subject headings. Relevant listings under "Telecommunications," "Satellites," "Cable Television," "Communication Policy," "Telephones," etc., cross reference bibliography.

0936   **PASCAL: BIBLIOGRAPHIE INTERNATIONALE: INTERNATIONAL BIBLIOG-RAPHY.** Paris: Institut National des Techniques de la Documentation, and Gesellschaft fur Mathematik und Datenverarbeitung (GFR), 1991–date, 10/year. Offers extensive non-English language journal coverage on information science literaure. Classified listings under "Communication, Information Transfer," "Telecommunications, Telematics." Author indexes; abstracts and subject indexes in both French and English.

0937   **TRANSDEX INDEX.** Ann Arbor, MI: UMI, 1975–date, monthly with annual cumulation in microfiche. Provides relatively timely key-word access to "non-U.S. originated materials selected and translated" by U.S. government's Joint Publications Research Service (JPRS). Extends annually to nearly 100,000 pages of translations; covers newspapers, trade and popular journals, speeches, and broadcasts in agriculture, business and economics, culture, politics, and science and technology. Materials in telecommunication are specifically selected: many are designated "TTP" documents;

other telecommunication documents appear in area designations (for Western Europe, China, etc.). Useful separate TTP "Source" list identifies publications of national telecommunication administrations. Keyed to TRANSDEX MICROFICHE collection with full-text translations.

7-A-3. Directories and Yearbooks

0938   Electronic Industries Association. **INTERNATIONAL ELECTRONICS CONTACTS**. Washington: Electronic Industries Association, 1992, 67 pp. Directory of information sources in 70 countries.

0939   **GLOBAL ECONOMIC CO-OPERATION: A GUIDE TO AGREEMENTS AND ORGANIZATIONS**. Bernard Colas, editor. Didcot, Oxfordshire: Management Books; Deventer, Netherlands, and Cambridge, MA: Kluwer; Tokyo: United Nations University Press, 1994 (2nd ed.), 557 pp. 1990 (1st ed.) in French. Descriptions of international organizations and their instruments (agreements, conventions, constitutions, regulations, etc.). "Communications" chapter covers telecommunication and satellites (pp. 306–319). Gives up-to-date information on governance, organization, and structure of ITU, Intelsat, Inmarsat, and Intersputnik. Chapters on trade, transnational corporations, standardization, industrial copyright, and use of outer space include useful information on governance of WIPO, GATT, UNCTAD, and others.

0940   **INFORMATION TECHNOLOGY ATLAS—EUROPE**. Amsterdam: IOS; London: C.G. Wedgwood, 1990, (2nd ed. revised and expanded), 451 pp. An "atlas" only in the sense that describes telecommunications in the European landscape; really a useful comprehensive directory of European organizations active in information and communication technology research, standardization, and infrastructure. Chapters describe research policies and programs, institutes and university programs, trade associations, IT companies (some 1,300 in number), standards bodies, PTTs and TOs, and professional organizations. Organizations and acronyms indexes.

0941   Schiavone, Giuseppe. **INTERNATIONAL ORGANIZATIONS: A DICTIONARY AND DIRECTORY**. London: Macmillan, 1992 (3rd ed.), 337 pp. Concise narrative descriptions of major organizations, including ITU and UN organizations. Separate listing of major "Posts and Telecommunications" organizations (pp. 294–95). Useful organizational membership table (identifying countries in ITU, for example); indexes for founding dates, acronyms, subjects, etc.

0942   Union of International Associations. **YEARBOOK OF INTERNATIONAL ORGANIZATIONS**. Munchen: K.G. Saur, 1948–date, annual. The most comprehensive—and physically complex—guide to international organizations. 30th ed. (1993) covers 1993/1994 in three massive volumes: vol. 1 includes descriptive data; vol. 2 arranges organization by countries; vol. 3 provides guide to subject indexing categories and separate subject and keyword indexes. Relevant listings under "Transportation, Telecommunications," and subdivisions including "Radio," "Satellites," "Telecommunications," "Telegraphs," "Telephones," and others.

7-B. International Telecommunication Union

7-B-1. Finding Aids for the ITU

0943   **CATALOGUE DES FILMS SUR LES TELECOMMUNICATIONS ET L'ELECTRONIQUE: CATALOGUE OF FILMS ON TELECOMMUNICATIONS AND ELECTRONICS: CATALOGO DE PELICULAS SOBRE TELECOMMUNICACIONES Y ELECTRONICA**. Geneva, Switzerland: ITU Central Library, 1971–date, biennial. Lists more than 500 films and videos in all fields of telecommunication and electronics held by the Film Library in the Central Library, Documentation and Archives Section and that are available for loan. Indexed by subject, title, language, country. Available free on request.

0944   El-Zanati, A. G. **DISSEMINATION OF INFORMATION AND DOCUMENTATION AT THE INTERNATIONAL TELECOMMUNICATION UNION IN THE '90s: A PAPER**

PRESENTED TO THE CEPT WORKING GROUP T/GT 17 "LIBRARY AND DOCUMEN-TATION" (GRONINGEN, THE NETHERLANDS, 13–15 JUNE 1990). Geneva, Swtizerland: ITU, 1990, 104pp. Useful for concise descriptions of function and method of operation of ITU, emphasizing its information dissemination activities. Details ITU, its major publications, TIES electronic information system available over Internet, and libraries.

See 0042 GLOSSARY OF TELECOMMUNICATIONS TERMS: ENGLISH, ARABIC, FRENCH, SPANISH.

0945　　LIST OF PUBLICATIONS. Geneva, Switzerland: ITU, 1979–date, semiannual. Bibliographic and ordering information for ITU publications. New publications through 1993 were also listed in TELECOMMUNICATION JOURNAL (1152).

0946　　LIST OF RECENT ACQUISITIONS/LISTE DES ACQUISITIONS RECENTES/ LISTA DE ADQUISICIONES RECIENTES. ITU Central Library, 1972–date, quarterly. Available free on request.

0947　　LIST OF ANNUALS/LISTE DES PUBLICATIONS ANNUELLES/LISTA DE PUBLICACIONES ANUALES. ITU Central Library, 1972–date, biennial. Nearly 400 titles. Available free on request.

0948　　LIST OF PERIODICALS/LISTE DES PERIODIQUES/LISTA DE REVISTAS. Geneva, Switzerland: ITU Central Library, 1967–date, biennial. More than 1,000 titles. Available free on request.

0949　　ITU GLOBAL DIRECTORY/REPERTOIRE GENERAL DE L'UIT/GUIA GENERAL DE LA UIT. 6th ed. Geneva, Switzerland: ITU, 1993, 372pp. "Official ITU desk reference of telecommunication officials and organizations": contacts and electronic and postal addresses for membership. English, French, Spanish editions; also available on ITU's TIES online information service.

7-B-2. ITU and Its Offices

0950　　International Telecommunication Union, Place des Nations, CH-1211, Geneva 20, Switzerland. Voice 41 22 730 5111. Fax 41 22 733 7256. Maintains TIES online information service. Internet: helpdesk@itu.ch. Created in 1865 as the International Telegraph Union, and broadened in the early 1900s to include telephone and wireless services, the ITU (a United Nations specialized agency since 1949) is the most important world-wide body in the field. In it reside (by both treaty and tradition) world-wide cooperation in allocation of radio frequencies and telecommunication technical standards. The first major reorganization of the ITU since 1947 took effect in mid-1994. Official organ is the monthly TELECOMMUNICATION JOURNAL (1152), now NEWSLETTER OF THE INTERNATIONAL TELECOMMUNICATION UNION (1147). See also its LIST OF PUBLICATIONS (0945) issued twice a year which details the many multi-volume official documents growing out of international meetings on telegraph-telephone, and on radiocommunication.

　　　1　Plenipotentiary Conference. All members of the ITU; convenes every 4 years to determine policies, approve budget, adopt proposals for amendments to the ITU's governance agreements (the Constitution and the Convention), conclude agreements with other international organizations, and elect members to the Council, Secretary-General, Radio Regulation Board, and other offices

　　　2 Secretary-General. The chief administrative officer of the ITU, in charge of the General Secretariat of the ITU.

　　　3 Council. Acts as ITU's governing body between plenipotentiary conferences (composed of 43 members). Exercises financial control over Secretary-General and sectors.

　　　4 Radiocommunication Sector. Responsible for use of radio frequency spectrum. Sponsors conferences, studies, study groups. Oversees Radio Regulation Board (replaces International Frequency Registration Board) and Radiocommunication Bureau.

　　　5 Telecommunication Standardization Sector. Responsible for recommending telecommunications standards world-wide. Studies technical, operating, and tariff issues; sponsors standardization conferences and study groups. Oversees Telecommunication Standardization Bureau.

6 **Telecommunication Development Sector**. Coordinates and provides technical assistance programs. Sponsors study groups. Operates Telecommunications Development Bureau. Voice: 41 22 730 5115.

7-C. Other Agencies of the United Nations

7-C-1. Finding Aids for the UN

0951   **DIRECTORY OF SELECTED COLLECTIONS OF UNITED NATIONS SYSTEM PUBLICATIONS**. New York: United Nations, 1991, 126 pp. "Compiled by the Advisory Committee for the Coordination of Information Systems (ACCIS)." International directory of UN depository libraries and information centers, with notes on scope of collections by document type (GATT, UNCTAD, WIPO, Unesco, etc.).

0952   **DIRECTORY OF UNITED NATIONS DATABASES AND INFORMATION SERVICES**. New York: United Nations, 1990 (4th ed.), 484 pp. "Compiled by the Advisory Committee for the Coordination of Information Systems (ACCIS)." Classified directory of 872 databases and information services available from 39 UN bodies as well as its specialized and related organizations, including Unesco, ITU, WIPO, GATT, etc. Subject index.

0953   **UNDOC: CURRENT INDEX: UNITED NATIONS DOCUMENTS INDEX**. New York: United Nations, 1979–date, monthly (10/year) with annual cumulation. CD-ROM version INDEX TO UNITED NATIONS DOCUMENTS AND PUBLICATIONS available from NewsBank/Readex: covers 1990–date, updated annually. Part I, arranged by agency, gives full bibliographic information; includes indexes for personal and corporate names and titles. Part II is subject index; cross references by unique sequential numbers. Useful headings include "Communication media," "Radio broadcasting," "Remote sensing," "Space communication," "Telecommunications," and "Telephone services." UNDOC does not index ITU publications.

0954   Unesco. **ANNOTATED BIBLIOGRAPHY OF UNESCO PUBLICATIONS AND DOCUMENTS DEALING WITH SPACE COMMUNICATION, 1953–1977**. Paris: Unesco, 1977, 102 pp. Some 281 annotated items, with title index.

0955   _____. **LIST OF DOCUMENTS AND PUBLICATIONS IN THE FIELD OF MASS COMMUNICATIONS**. Paris: Unesco, 1976–date, annual. Usually one volume per year; includes increasing number of telecommunication references despite title.

0956   **YEARBOOK OF THE UNITED NATIONS**. New York: United Nations, Department of Public Information; Dordrecht, The Netherlands: Martinus Nijhoff, 1946–date, annual. Official review of UN's work in topical and regional chapters: "Political and security questions," "Economic and social questions," for example. Also includes brief reports of UN's intergovernmental organizations, such as ITU, Unesco, WIPO, GATT, etc. Subject index references relevant information under "Satellites," "Telecommunications," "Transport/Communications," and similar subdivisions under regions and countries.

7-C-2. Selected UN Agencies

0957   **Advisory Committee for the Coordination of Information Systems** (ACCIS). ACCIS Secretariat, Palais des Nations, 1211 Geneva 10, Switzerland. Voice: 41 22 798 8591. Fax: 41 22 740 1269. Facilitates member access to UN information and promotes improvement of information infrastructure within UN system. Compiles DIRECTORY OF UNITED NATIONS DATABASES AND INFORMATION SERVICES (0952) and DATABASE OF UNITED NATIONS SERIAL PUBLICATIONS (1988), and others.

0958   International Civil Aviation Organization (ICAO). 1000 Sherbrooke St. West, Montreal, Quebec H3A 2R2, Canada. Voice: 1 514 285 8219. Fax: 1 514 288 4772. Maintains databases of international air navigation communication frequencies.

0959   International Maritime Organization (IMO). 4 Albert Embankment, London SE1 7SR, United Kingdom. Voice: 44 71 735 7611. Fax: 44 71 587 3210. Maritime radio and peaceful uses of space.

0960   United Nations Conference on Trade and Development (UNCTAD). Palais des Nations, 1211 Geneva 10, Switzerland. Voice: 41 22 734 6011. Fax: 41 22 733 9879. Promotes international trade to accelerate economic development. Particular interest in technology transfer and role of technology in economic development. UNCTAD's International Trade Centre (ITC) promotes technical cooperation with developing countries in trade promotion. In process of creating global network of trading centers and information clearinghouses electronically linking buyers, sellers, banks, etc., some 47 "trade points" to date: North American trade point in Columbus, OH. Holds World Summit on Trade Efficiency. Publishes TRADE AND DEVELOPMENT: AN UNCTAD REVIEW (1979–date).

0961   United Nations Development Programme (UNDP). One United Nations Plaza, New York, NY 10017. Voice: 212 906 5000. Fax: 212 826 2058. Supports communication projects of national governments in Asia, Africa, Latin America, Arab States, and parts of Europe and with international agencies to promote higher standards of living and faster economic growth.

0962   United Nations Educational, Scientific, and Cultural Organization (Unesco). 7 Place de Fontenoy, 75700 Paris, France. Voice: 33 1 4568 1000. Fax: 31 1 4567 1690. Promotes world peace and security through mass communication and knowledge diffusion. Publishes LIST OF DOCUMENTS AND PUBLICATIONS IN THE FIELD OF MASS COMMUNICATIONS (0955) and WORLD COMMUNICATION REPORT (0929) and sponsors International Programme for the Development of Communication, which publishes IPDC NEWSLETTER (1141) and other publications. Maintains Culture and Communication Documentation Centre.

0963   United Nations Industrial Development Organization (UNIDO). Vienna International Centre, Wagramerstrasse 5, P.O. Box 300, 1400 Vienna, Austria. Voice: 43 222 211 31. Fax: 43 222 232 156. Promotes and assists developing countries toward industrialization. Coauthored with ESCAP, TRANSPORT AND COMMUNICATIONS DECADE FOR ASIA AND THE PACIFIC (1992).

0964   UN Regional Economic Commissions:
          1 **Economic and Social Commission for Asia and the Pacific** (ESCAP). United Nations Bldg., Rajadamnern Ave., Bangkok 10200, Thailand. Voice: 66 2 282 9161. Fax: 66 2 282 9602. Initiates and promotes social and economic development programs in Asia and the Pacific, including in fields of communications. ESCAP's Transportation and Communication Division publishes TECHNOLOGY INFORMATION ORGANIZATIONS IN ASIA AND THE PACIFIC, and TRANSPORT AND COMMUNICATIONS BULLETIN FOR ASIA AND THE PACIFIC among others.
          2 **Economic Commission for Africa** (ECA). P.O. Box 3011, Addis Ababa, Ethiopia. Voice: 251 1 51 7200. Fax: 251 1 51 4416. Promotes social and economic development in Africa. Maintains ministerial-level Conference of African Ministers of Transport, Communications, and Planning.
          3 **Economic Commission for Europe** (ECE). Palais des Nations, 1211 Geneva 10, Switzerland. Voice: 41 22 734 6011. Fax: 41 22 733 9879. Promotes intergovernment cooperation among countries of Europe and United States and Canada. ECE has direct responsibility for programs dealing with standardization, automation, and engineering, especially development of Electronic Data Interchange for Administration, Commerce, and Transport (EDIFACT). Published THE TELECOMMUNICATIONS INDUSTRY: GROWTH AND STRUCTURAL CHANGE (1987), a detailed examination of the European telecommunications sector focusing on industry integration and convergence and joint ventures for international competitiveness.

4 Economic Commission for Latin America and the Caribbean (ECLAC). Casilla 179-D, Santiago, Chile. Voice: 56 2 48 5051. Fax: 56 2 48 0252. Operates library with collections covering copyright, industrial design, industrial property, patents, trademarks, and unfair competition. Promotes raising level of economic activity in region and inter-regional cooperation; sponsors research on economic and technological problems.

0965 **United Nations University** (UNU). Toho Seimei Bldg., 29th Floor, 15-1 Shibuya 2-Chome, Tokyo 150, Japan. Voice: 81 3 499 2811. Fax: 81 3 499 2828. Email: (Bitnet) dellata@jpnunuoo. Promotes development through distance education. Produces BIBLIOGRAPHIC DATABASE ON THE LITERATURE OF DISTANCE EDUCATION. Maintains International Centre for Distance Learning (ICDL), Open University, Walton Hall, Milton Keynes MK7 6AA, U.K.; internet kw_harry@uk.ac.open.acsvax.

0966 **World Intellectual Property Organization** (WIPO). 34 Chemin des Colombettes, 1211 Geneva 20, Switzerland. Voice: 41 22 730 9111. Fax: 41 22 733 5428. Promotes intellectual property rights and protection. Operates library with collections covering copyright, industrial design, industrial property, patents, trademarks, and unfair competition.

7-D. International Satellites

7-D-1. INTELSAT

0967 **International Telecommunications Satellite Organization** (INTELSAT). 3400 International Dr. NW, Washington, DC 20008. Voice: 202 944 6800; 944 7500. Fax: 202 944 7890. Members include more than 120 governments. Owns and operates global satellite system. Sponsors programs that provide advice and financial support for planning or improving telecommunications systems and training. Technical and research offices in California. Publishes reports, bibliographies (0910 and 0911), INTELSAT ANNUAL REPORT (1974–date, annual), and INTELSAT NEWS (1985–date, quarterly).

7-D-2. Directories of International Satellites

0968 Martin, Donald H. **COMMUNICATION SATELLITES, 1958–1992**. El Segundo, CA: Aerospace Corporation, 1991, 368 pp. Chapters cover historical background and give data (configuration, capacity, transmitter, receiver, antenna, design, orbit, and operations and management) for experimental, international, military, Soviet, Canadian-U.S., European, domestic and regional, and other (engineering, testing, educational) satellites. Entries include bibliographies. Glossary and extensive classified bibliography appendixed.

0969 Pelton, Joseph N., and John Hawkins, editors. **SATELLITES INTERNATIONAL**. New York: Stockton Press, 1987, 265 pp. Lists communication satellites and types of application satellites worldwide. Articles, information on manufacturers, launch vehicles, global and regional systems. Listing of satellites by country.

0970 **SPACE ACTIVITIES OF THE UNITED NATIONS AND INTERNATIONAL ORGANIZATIONS**. New York: United Nations, 1992, 318 pp. Subtitle: "A review of the activities and resources of the United Nations, its specialized agencies and other international bodies relating to the peaceful uses of outer space." Covers ITU (pp. 93–109), International Maritime Organization (pp. 128–32), WIPO (pp. 132–34), as well as European Space Agency, INTELSAT, INTERSPUTNIK, INMARSAT, EUTELSAT, Arab Satellite Communication Organization, and others. Bibliography of UN documents related to outer space (pp. 281–318).

0971 **WORLD SATELLITE DIRECTORY**. Potomac, MD: Phillips, 1979-date, annual. 15th ed. (1993) profiles more than 5,000 domestic and international satellite operators, transponder brokers and sellers, product manufacturers and suppliers and professional services, and U.S. and foreign regulators, government agencies, and administrations. Product and company indexes. Glossary, table of abbreviations.

0972   Stanyard, Roger. **WORLD SATELLITE SURVEY.** London: Corp. of Lloyd's, 1987, 375 pp. Directory of 92 telecommunication systems.

7-D-3. *Other Regional Satellite Organizations*

0973   **Arab Satellite Communications Organization** (ARABSAT). P.O. Box 1038, Riyadh, Saudi Arabia. Voice: 966 1 4646 666. Operates regional system for television, telephony, and mobile and data communications.

0974   **European Telecommunications Satellite Organization** (EUTELSAT). Tour Maine Montparnasse, 33 Avenue du Maine, 75755 Paris, France. Voice: 33 1 4538 4747. Fax: 33 1 4538 3700. Includes 26 national telecommunications administrations of western Europe.

0975   **International Maritime Satellite Organization** (INMARSAT). 40 Melton St., London NW1 2EQ, United Kingdom. Voice: 387 9089. Fax: 387 2115. Members include 59 countries.

7-D-4. *Policy and Regulation*

0976   Bockstiegel, Karl-Heinz, and Marietta Benko. **SPACE LAW: BASIC LEGAL DOCUMENTS.** Dordrecht, The Netherlands: Martinus Nijhoff, 1990, two loose-leaf vols. Texts of major legal instruments (vol. 1) and comment on same (vol. 2) including telecommunications, access to the geostationary orbit, major international bodies, national legislation from different countries, etc.

0977   Demac, Donna A., ed. **TRACING NEW ORBITS: COOPERATION & COMPETITION IN GLOBAL SATELLITE DEVELOPMENT.** New York: Columbia University Press, 1986, 329 pp. Eighteen papers on the emergence of specialized and regional satellite systems, national satellite programs, regulation of international satellite activity, and Soviet satellites. Bibliography, notes, index.

0978   Martinez, Larry. **COMMUNICATION SATELLITES: POWER POLITICS IN SPACE.** Dedham, MA: Artech, 1985, 186 pp. International political issues in obtaining orbital slots and the necessary uplinks and downlinks for service. Glossary, bibliography.

0979   Smith, Milton L. **INTERNATIONAL REGULATION OF SATELLITE COMMUNI-CATION.** Dordrecht, The Netherlands: Nijhoff, 1990, 245 pp. Institutional framework, ITU space WARC meetings, and the new regulatory regime for the fixed satellite service. Notes.

0980   Snow, Marcellus S. **THE INTERNATIONAL TELECOMMUNICATIONS SATELLITE ORGANIZATION (INTELSAT): ECONOMIC AND INSTITUTIONAL CHALLENGES FACING AN INTERNATIONAL ORGANIZATION.** Baden-Baden, Germany: Nomos Verlagsge-sellschaft, 1987, 200 pp. Includes discussion of separate satellite systems. Appendices, notes.

0981   White, Rita Lauria and Harold M. White, Jr. **THE LAW AND REGULATION OF INTER-NATIONAL SPACE COMMUNICATION.** Norwood, MA: Artech, 1988. Well-organized overview showing domestic and international overlaps and relationships.

7-E. *Other Telecommunications Entities*

7-E-1. *Selected International Entities*

0982   **General Agreement on Tariffs and Trade** (GATT). Centre William-Rappard, 154 Rue de Lausanne, 1211 Geneva 21, Switzerland. Voice: 41 22 739 5111. Fax: 41 22 731 4206. Provides a forum for the development of rules and agreements to promote the liberalization of trade. Linked to the ITU (0950) and UN regional organizations.

0983   **International Standards Organization** (ISO). 1, rue de Varembe, Case Postale 56, CH-1211, Geneva, 20 Switzerland. Voice: 41 22 749 0111. Fax: 41 22 733 3430. Issues ISO CATALOGUE (0403).

7-E-2. Selected Regional Entities

0984 **African Posts and Telecommunications Union** [Union Africaine des Postes et Telecommunications] (UAPT). Avenue Patrice Lumumba, P.O. Box 44, Brazzaville, Zaire. Voice: 242 83 2778. Represents postal and telecommunications corporations in French-speaking countries. Focuses on tariff research, regulation, and training. Publishes REVUE UAPT (1978-date, irregular).

0985 **Arab Telecommunication Union** (ATU). P.O. Box 2397, Baghdad, Iraq. Voice: 964 1 776 0901. Promotes improvement of telecommunications between members. Sponsors research. Publishes ARAB TELECOMMUNICATIONS UNION JOURNAL (1960-date).

0986 **Asia-Pacific Telecommunity** (APT). No. 12/49, Soi 5, Chaengwattaria Rd., Thungsanghong, Bangkhen, Bangkok 10210, Thailand. Voice: 66 2 573 6893. Fax: 66 2 573 7479. Promotes telecommunication development in ESCAP region. Focuses on planning and operating national and international networks. Provides assistance and training programs.

0987 **Association of State Telecommunications Undertakings of the Andean Subregional Agreement.** P.O. Box 17-1106042, Quito, Ecuador. Voice: 593 2 54 99 61. Sponsors research and concludes agreements to promote inter-regional telecommunications development.

0988 **Caribbean Association of National Telecommunications Organizations** (CANTO). c/o Trinadad-Tobago Telephone Company, 116 Frederich St., P.O. Box 917, Port-of-Spain, Trinadad-Tobago. Voice: 1 809 624 3081. Fax: 1 809 624 2268. Some 23 member countries. Publishes CANTO QUARTERLY REVIEW (1985–date), directories, and reports.

0989 **Caribbean Telecommunication Union** (CTU). 17 Queen's Park West, Port-of-Spain, Trinadad-Tobago. Voice: 1 809 628 8037. Fax: 1 809 628 6037. Promotes development of intra-regional, regional, and international telecommunications and cooperation with international organizations.

0990 **Commonwealth Telecommunications Organization.** Clareville House, 26–27 Oxendon St., London SW1Y 4EL, United Kingdom. Voice: 44 71 930 5516. Fax: 44 71 930 4248. Supports telecommunications through collaborative and cooperative arrangements and assistance of 31 members. Cooperates with other international organizations.

0991 **Conference des Administrations des Postes et Telecommunications de l'Afrique Centrale** (CAPTAC). BP 728, Yaounde, Cameroon. Voice: 237 20 48 54. Fax: 237 21 32 93. Promotes and develops posts and telecommunications services in 7 member countries. Cooperates with international organizations.

0992 **Conference of Southern African Telecommunications Administrations** (SATA). BP 2677, Maputo, Mozambique. Voice: 258 202 46. Consists of telecommunications administrations of 10 member countries. Cooperates with international organizations.

0993 **European Conference of Postal and Telecommunications Administrations** [Conference Europeenne des Administrations des Postes et des Telecommunications (CEPT). Seilerstrasse 22, CP 1283, CH-3001, Berne, Switzerland. Voice: 41 31 62 20 79. Fax: 41 31 62 20 78. Subsidiary of the ITU consisting of western European postal and telecommunications administrations. Active in telecommunications standardization and harmonization.

0994 **European Telecommunications Standards Institute** (ETSI). F-06921 Sophia Antipolis, Cedex, France. Voice: 33 92 94 42 00. Fax: 33 93 65 47 16. Enforces standards and specifications development system established by the European Union.

0995 **Inter-American Telecommunication Conference** (CITEL). c/o OAS, 1889 F St. NW, Washington 20006. Voice: 202 458 3004. Fax: 202 458 3967. Promotes telecommunications development in western hemisphere through training programs and technical meetings.

0996 **Panafrican Telecommunication Union** (PATU). c/o UNDP, P.O. Box 7248, Kinshasa, Zaire. Voice: 243 12 22175. Responsible for implementation and operation of Pan-African Telecommunications Network (PANAFTEL). Focuses on regional cooperation and development of African telecommunication equipment manufacturing industry.

0997   **South Pacific Telecommunications Development Programme** (SPTDP). GPO Box 856, Ratu Sakuna Rd., Suva, Fiji. Voice: 679 312 600. Fax: 679 302 203. Implements training and financial development programs under sponsorship of South Pacific Bureau for Economic Cooperation.

7-F. Statistical Sources for International Telecommunications Trade

0998   Department of Commerce. **NATIONAL TRADE DATA BANK**. [CD-ROM]. Washington, DC: GPO, 1990–date, monthly. Data on U.S. and international trade and trade conditions worldwide. Sources include Bureaus of Census, Bureau of Economic Analysis, International Trade Administration, departments of Labor and Commerce, Office of U.S. Trade Representative, U.S. Export-Import Bank, and more.

0999   Electronic Industries Association. **ELECTRONIC FOREIGN TRADE**. Washington, DC: EIA, 1982–date, monthly. Import and export figures for 300 products, by unit sales and dollar volume; balance of trade by product group; rankings of markets and suppliers. Based largely on U.S. Commerce Department data. Indexed in STATISTICAL MASTERFILE (0513).

1000   **INTERNATIONAL DIRECTORY OF TELECOMMUNICATIONS: MARKET TRENDS, COMPANIES, STATISTICS, AND PERSONNEL**. London: Longman, 1984–date, irregular. Latest ed. 1990 (4th ed.). Arranged by region and country: descriptions of national telecommunications infrastructures, covering major operators and regulatory agencies, networks, services, manufacturing industry, and international trade, with directories. Data for telephones and telex units, lines, revenues, etc. Glossary.

1001   Organization for Economic Cooperation and Development. **COMMUNICATIONS OUTLOOK**. Paris: OECD Publications and Information Center, 1990–date, biannual. International coverage of year's telecommunications activities. Charts, graphs, maps, and tables supplement brief summary assessments of trends around world and in major regions, emphasizing Europe.

1002   **U.S. INTERNATIONAL TELECOMMUNICATION RATE HISTORY**. Washington, DC: GPO (Dept. of Commerce Office of Telecommunications Report 75–77), 1975, 197 pp. Useful historical data (over about a decade) on rates for varied kinds of service between the United States and other countries. Appears not to have been updated, at least not in this form.

See 0929. **WORLD COMMUNICATION REPORT**.

1003   **WORLD TELECOM DATABOOK**. Boxgrove, Chichester, West Sussex: Telecom Information Services, 1991, 630 pp. Arranged by region and country: provides tables of basic data, including subscriber and leased lines, telephone penetration, digital lines, payphones, fax machines, mobile subscribers, etc. Market overviews, descriptions of regulatory and operator structures, telecommunications equipment markets, with directory of contacts.

1004   **THE WORLD'S TELEPHONES: A STATISTICAL COMPILATION**. New York; later Morris Plains, NJ: AT&T Communications, 1912–date, annual (not published for 1915–1918; 1942–45). Provides data for January 1st each year in tables and (in later years) charts and graphs. Includes data by region, country, and (in more recent years) major world cities, and telephones per 100 persons. Formerly TELEPHONE STATISTICS OF THE WORLD (through 1955).

1005   **YEARBOOK OF COMMON CARRIER TELECOMMUNICATION STATISTICS**. Geneva: International Telecommunication Union, 1974–date, annual. Alternate titles include ANNUAIRE STATISTIQUE DES TELECOMMUNICATIONS DU SECTEUR PUBLIC ET STATISTIQUES DES TELECOMMUNICATIONS. Based on official reports from ITU member countries, each volume reports data for the previous decade, including telephone and other service penetration, investment, etc. Indexed in SATISTICAL MASTERFILE (0513).

7-G. Selected Secondary Resources

7-G-1. Broad Surveys

1006   Akwule, Raymond. **GLOBAL TELECOMMUNICATIONS: THE TECHNOLOGY, ADMINISTRATION, AND POLICIES**. Boston: Focal Press, 1992, 200 pp. A very brief survey. Tables, index.

1007   Branscomb, Anne W., ed. **TOWARD A LAW OF GLOBAL COMMUNICATIONS NETWORKS**. White Plains, NY: Longman, 1986, 370 pp. Twenty-seven chapters, most authored by attorneys, discuss the global telecommunications environment, information users and information transfer systems, and emerging legal and policy issues in global information flow. Bibliography, index.

1008   Bruce, Robert B., et al. **THE TELECOM MOSAIC: ASSEMBLING THE NEW INTERNATIONAL STRUCTURE**. Stoneham, MA: Butterworths, 1988, 447 pp. Useful record of early liberalization in Europe and elsewhere with sections on value-added services, telecommunications and transactional services, the future of European policies in this sector, the changing environment for planning international facilities, and telecommunication structures in the developing world.

1009   Cherry, Colin. **WORLD COMMUNICATION: THREAT OR PROMISE? A SOCIO-TECHNICAL APPROACH** New York: Wiley, 1978 (2nd ed.), 229 pp. Changing influence of communication technology around the world is assessed. Tables, charts, bibliography, index.

1010   Comor, Edward A., ed. **THE GLOBAL POLITICAL ECONOMY OF COMMUNICATION: HEGEMONY, TELECOMMUNICATION, AND THE INFORMATION ECONOMY**. New York: St. Martin's, 1994, 193 pp. Nine essays by academics and researchers on implications of global telecommunications industrial development on U.S. and other national political structures. Index.

1011   Crandall, Robert W., and Kenneth Flamm, eds. **CHANGING THE RULES: TECHNOLOGICAL CHANGE, INTERNATIONAL COMPETITION, AND REGULATION IN COMMUNICATIONS**. Washington, DC: Brookings, 1989, 424 pp. Ten largely economic papers, most of them focused on the U.S., but a few placing this country in global context. Tables, charts, index.

1012   Eward, Ronald. **THE DEREGULATION OF INTERNATIONAL TELECOMMUNICATIONS**. Dedham, MA: Artech House, 1985, 425 pp. The overseas marketplace, FCC actions and their effect, evolving technology, global networks, international facility planning, long-term market positions of major carriers, and international deregulation. Tables, charts, references.

1013   Fortner, Robert S. **INTERNATIONAL COMMUNICATION: HISTORY, CONFLICT, AND CONTROL OF THE GLOBAL METROPOLIS**. Belmont, CA: Wadsworth Publishing, 1993, 390 pp. Textbook examining history and development of international mass media systems largely in a political context from 1835 to date. Particular focuses include international information flow, imperialism and intellectual property, human rights, social control and propaganda, and law and regulation, with selective coverage of technological issues. Appendixes, glossary, bibliography, index.

1014   Frederick, Howard H. **GLOBAL COMMUNICATION AND INTERNATIONAL RELATIONS**. Belmont, CA: Wadsworth, 1993, 287 pp. Discusses the means and uses of international communication (media and telecommunication), contending theories of international communication, communication in war and peace, and global communication law. Tables, notes, index.

1015   Harasim, Linda M., ed. **GLOBAL NETWORKS: COMPUTERS AND INTERNATIONAL COMMUNICATION**. Cambridge, MA: MIT Press, 1993, 411 pp. Twenty-one readings suggest new and evolving applications of computer networking in global communication. Emphasis is on the widening abilities of individuals with personal computers and modems to communicate world-wide. Index.

1016   Jussawalla, Meheroo, and Chee-Wah Cheah, **THE CALCULUS OF INTERNATIONAL COMMUNICATIONS: A STUDY IN THE POLITICAL ECONOMY OF TRANSBORDER DATA FLOWS**. Littleton, CO: Libraries Unlimited, 1987, 159 pp. The history and nature of such flows, economic aspects of policy in this area, and prospects for TBDF policy coordination. Notes, references, index.

1017   Jussawalla, Meheroo, et al., eds. **INFORMATION TECHNOLOGY AND GLOBAL INTERDEPENDENCE**. Westport, CT: Greenwood, 1989, 321 pp. See next entry.

1018   Jussawalla, Meheroo, ed. **GLOBAL TELECOMMUNICATIONS POLICIES: THE CHALLENGE OF CHANGE**. Westport, CT: Greenwood Press, 1993, 280 pp. Two useful collections of scholarly papers on different aspects of international communication infrastructure amidst dramatic change. Index.

1019   Katz, Raul Luciano. **THE INFORMATION SOCIETY: AN INTERNATIONAL PERSPECTIVE**. New York: Praeger, 1988, 168 pp. Means of measuring the global information workplace and workforce, and the impact of government policies. Tables, figures, bibliography, index.

1020   Mansell, Robin. **THE NEW TELECOMMUNICATIONS: A POLITICAL ECONOMY OF NETWORK EVOLUTION**. London and Thousand Oaks, CA: Sage, 1993, 120 pp. Useful and often critical assessment of network developmental policies (both government and private sector) in the U.S., Britain, France, Germany, and Sweden as well as American experiences. Author draws overall conclusions to both describe and critique world patterns. Glossary, bibliography, index.

1021   Savage, James G. **THE POLITICS OF INTERNATIONAL TELECOMMUNICATIONS REGULATION**. Boulder, CO: Westview, 1989, 240 pp. The ITU, radio frequency spectrum management, free flow vs. prior consent issues, and international technical standards. Notes, index.

1022   Stevenson, Robert L. **GLOBAL COMMUNICATION IN THE TWENTY- FIRST CENTURY**. White Plains, NY: Longman, 1994, 382 pp. An overall text survey with much on media, but several chapters detailing broader telecommunication concerns. Index.

1023   Wellenius, Bjorn and Peter A. Stern, eds. **IMPLEMENTING REFORMS IN THE TELECOMMUNICATIONS SECTOR: LESSONS FROM EXPERIENCE.** Washington: World Bank, 1994, 757 pp. Forty papers, most original to this volume discuss the current state of telecommunications regulation and structure world-wide, recent experience in Latin America, recent experience in the Asia-Pacific region, recent experiences in Europe, foreign operators' perspective on privatization, mobilizing capital for privatization, issues of regulation, and a conclusion on the strategic issues of implementation. Notes, glossary, annex on current sector structure and regulatory framework in selected countries and regions (90 pages).

7-G-2. International Telecommunication Management

1024   Bernt, Phyllis, and Martin Weiss. **INTERNATIONAL TELECOMMUNICATIONS**. Carmel, CA: Sams Publishing, 1993, 466 pp. A text with many case studies covering regulation, economics, organization, and technologies.

1025   Elbert, Bruce R. **INTERNATIONAL TELECOMMUNICATION MANAGEMENT** Norwood, MA:Artech House, 1990, 343 pp. International network environment, circuit-switched and private line services, packet-switched services, private ownership of networks, and equipment, international transmission of audio and video, management and logistics, and regulation of transborder data flows. Charts, bibliography, index.

1026   **INFORMATION TECHNOLOGY AND INTERNATIONAL COMPETITIVENESS: THE CASE OF THE CONSTRUCTION SERVICES INDUSTRY**. New York: United Nations, 1993, 182 pp. Sponsored by UNCTAD. Case studies of information technology usage in U.S., U.K., and Sweden.

1027 Lanvin, Bruno. **INTERNATIONAL TRADE IN SERVICES, INFORMATION SERVICES, AND DEVELOPMENT.** Geneva, Switzerland: UNCTAD, 1987, 46 pp. (United Nations Conference on Trade and Development Discussion Papers, 23). Overview of transnational corporations and joint ventures in the telecommunications industry.

1028 Vignault, Walter L. **WORLDWIDE TELECOMMUNICATIONS GUIDE FOR THE BUSINESS MANAGER.** New York: Wiley, 1987, 417 pp. International information flow, network attachment products, office systems, digital voice and data networks, U.S. network services and alternatives, international traffic, satellite communications. Directories, bibliography, glossary, index.

7-G-3. U.S. Role Overseas

1029 Aronson, Jonathan David and Peter F. Cowhey. **WHEN COUNTRIES TALK: INTERNATIONAL TRADE IN TELECOMMUNICATIONS SERVICES.** Cambridge, MA: Ballinger, 1988, 289 pp. Impact of deregulatory policies here and abroad on such trade, the varied roles of competitive provision of services, four models of the world market, and the role of the GATT. Tables, notes, index.

1030 Barnett, Stephen R., et al. **LAW OF INTERNATIONAL TELECOMMUNICATIONS IN THE UNITED STATES.** Baden-Baden, Germany: Nomos Verlagsgesellschaft, 1988, 71 pp. Regulation of both American common carriers and electronic media in international service. Notes.

1031 Golden, James R. **ECONOMICS AND NATIONAL STRATEGY IN THE INFORMATION AGE: GLOBAL NETWORKS, TECHNOLOGY POLICY, AND COOPERATIVE COMPETITION.** Westport, CT: Praeger, 1994, 301 pp. Argues for "cooperative competition" (p. xv) in post-Cold War information age in which U.S. exploits advantages as strategic broker in international agreements governing technology development. Sees U.S. information technology policies converging with U.S. foreign trade and military policies. Bibliography, index, figures, and tables.

1032 National Telecommunications and Information Administration. **U.S. TELECOMMUNICATIONS IN A GLOBAL ECONOMY: COMPETITIVENESS AT A CROSSROADS.** Washington: GPO, 1990, 260 pp. Detailed study of factors in American competitiveness and trade imbalance in this sector. Tables.

1033 Noam, Eli, and Gerard Pogorel, ed. **ASYMMETRIC DEREGULATION: THE DYNAMICS OF TELECOMMUNICATIONS POLICY IN EUROPE AND THE UNITED STATES.** Norwood, NJ: Ablex, 1994, 264 pp. Ten essays by academics and researchers in fields of economics, law, political science on different regulatory, policy, trade, and pricing issues in Europe and U.S. Index.

1034 Office of Technology Assessment. **U.S. TELECOMMUNICATIONS SERVICES IN EUROPEAN MARKETS.** Washington: GPO, 1993, 220 pp. Roles of American government in helping encourage competitive policies, and American firms in actually offering service in the opening markets of Europe. Charts, diagrams, notes.

1035 Sterling, Christopher H., ed. **INTERNATIONAL TELECOMMUNICATIONS AND INFORMATION POLICY.** Washington, DC: Communications Press, 1984, 496 pp. Based on a day-long symposium, this includes discussions on the role of the U.S. in international communication, reshaping American government for the information age, domestic deregulation and international policies, and a reprint of the NTIA repport, "Long-range goals in international telecommunications and information." Abbreviations list.

1036 Thimm, Alfred L. **AMERICA'S STAKE IN EUROPEAN TELECOMMUNICATION POLICIES.** Westport, CT: Quorum/Greenwood, 1992, 254 pp. Assesses developments in Germany, France, Britain, and the European Community as they affect American policy and companies. Figures, tables, index.

8

Periodicals

Our final chapter describes the many telecommunications periodicals. We cover periodicals together—as they appear in many libraries—but here arranged in topical divisions. Here especially we note that our coverage is severely selective, including, for example, only English-language journals. Even more significantly, we only include journals that substantially focus on telecommunications. Among useful titles that we have intentionally excluded are most that focus on electronic media, because while they increasingly include telecommunications, that is not their primary role.

A far greater—and very conscious— omission is the multitude of telecommunications newsletters, most of which—aimed at markets—are priced beyond the reach of libraries and individual researchers and are ephemeral in form and frequently in quality, at least for authoritative longitudinal research. We have included the two that we and others regard as among the most important, the best values, and the most authoritative: COMMUNICATIONS DAILY (1037) and TELECOMMUNICATIONS REPORTS (1041). Full-text electronic versions of many telecommunications newsletters, however, are available in online and/or CD-ROM resources including LEXIS/NEXIS (0728), BUSINESS NEWS (0491), and F & S INDEX UNITED STATES (0493), and from Dialog, NewsNet, and other electronic services. TELECOMMUNICATIONS ALERT (1040) excerpts features from a wide selection of newsletters. Popovich and Costello's guide (0487) identifies a wide selection by service. The need for a complete comparative listing of telecommunications newsletters remains.

8-A. *General*

1037   **COMMUNICATIONS DAILY.** Washington: Warren Publishing, 1981–date, daily. Full-text available online on Dialog, LEXIS/NEXIS (0728), NewsNet. The most useful newsletter for information about telecommunications in the broadest sense. Features brief notices on business, legal, and technological activities relevant to the telephone, broadcasting, cable, video, and other information technologies.

1038   **NETWORK WORLD.** Framingham, MA: CW Communications, 1983–date, weekly. Full-text available online on LEXIS/NEXIS (0728). Current information on issues and developments in the data communication industries.

1039   **TELECOMMUNICATIONS.** Norwood, MA: Horizon House-Microwave, Inc., 1967–date, monthly. Full-text available online on Dialog. Current information relevant to industry, with particular emphasis on government regulation, industry trends, technological research, and new products. Authors typically employed in telecommunications. Most articles are technical in nature, but aimed at business and management (non-technical) professionals interested in competitive advantages. Feature stories on timely topics and "hottest technologies," such as ATM, wireless communications, ISDN development, computer viruses, facsimile transmission systems, metropolitan area newtorks (MANs), teleports, and fiber optics. Contributions by top industry managers and regulators. Informed interviews and editorials. Heavily illustrated. Special theme issues.

1040   **TELECOMMUNICATIONS ALERT.** New York: Management Telecommunications Publishing, 1983–date, monthly. Full-text available online on NewsNet, PTS Newsletter Database.

Digests news features (legislation, regulation, technical, business) selected from several hundred other major telecommunication trade journals and popular press, including COMMUNICATIONS DAILY (1037), FCC WEEK, and PERSONAL COMPUTING. Subject and company indexes. Most useful when full-text services like LEXIS/NEXIS (0728) and NewsNet are not available.

1041   **TELECOMMUNICATIONS REPORTS.** Washington: Business Research Publications, 1934–date, weekly. Full-text available online on NewsNet. Historically important trade newsletter. Emphasis largely on the FCC and regulation of telecommunication; tracks legislative, regulatory, and industry activities related to voice and data services. Selected articles reprinted in FIFTY YEARS OF TELECOMMUNICATIONS REPORTS (0311). Complements COMMUNICATIONS DAILY (1037).

8-B. History

1042   **TECHNOLOGY AND CULTURE.** Chicago: University of Chicago Press, 1959–date, quarterly. Scholarly journal cutting across all aspects of technology and its social impact. Often includes useful papers on communication.

1043   **HISTORY AND TECHNOLOGY.** Yverdon, Switzerland: Harwood Academic, 1983–date, quarterly. Scholarly articles on "technology as embodied in society, exploring its links between science, on the one hand, and the cultural, economic, political and institutional contexts of its genesis and development on the other." Particular emphasis on national and international technology and research and development policies. Recent special issues vol. 11, number 1 (1994), "Information Technologies and Socio-Technical Systems"; and vol. 11, number 2 (1994), "The Electronics Challenge: An Historical Perspective."

1044   **TELECOM HISTORY.** Dublin, CA: Telephone History Institute, 1994–date, twice a year. Avid amateur historians fill the pages of this recent journal with articles on early telephony. Each illustrated issue centers on a specific theme.

8-C. Technology

8-C-1. Telecommunications and Information Technology Research

1045   **AEU: ARCHIV FUR ELEKTRONIK UND UBERTRAGUNGSTECHNIK.** Stuttgart, Germany: S. Hirzel Verlag, 1947–date, bimonthly. International journal in communications, electronics, networks, information theory, antennas and propagation, optical communications, and other topics. Articles in English and German.

1046   **COMPUTER COMMUNICATIONS.** New York: IPC Science and Technology Press, 1978–date, bimonthly. International journal on computer communications, networks, and standards. Book, product, and literature reviews.

1047   **COMPUTER COMMUNICATIONS REVIEW.** New York: Association for Computing Machinery, 1970–date, quarterly. Journal of the "ACM Special Interest Group on Data Communications." Research articles on "data communication systems for computers, data communication technology for computers; reliability, security, and integrity of data in data communication systems; problems of interfacing communication systems and computer systems; computer communications systems modelling and analysis." Regular and occasional bibliographic features include "Bibliography of Recent Publications on Computer Communications" and "Annotated Bibliography on Network Management." Book reviews. Calls for papers, conference calendar and reports, announcements.

1048   **COMPUTER NETWORKS AND ISDN SYSTEMS: THE INTERNATIONAL JOURNAL OF COMPUTER AND TELECOMMUNICATIONS NETWORKING.** Amsterdam: Elsevier Science Publishers, 1977–date, 10/year in 2 vols. Formerly COMPUTER NETWORKS (1977–1984). Official journal of the International Council for Computer Communication. Includes highly technical articles on "all aspects of the design, implementation, use and management of computer and telecommunication networks, communication sub-systems, and integrated services digital networks" and "non-technical" articles on telecommunication's economic, legal, and regulatory issues, and social impact. Articles focus on ATM, OSI, LANs, computer viruses, bandwidth allocations, simulated analysis of protocols, conformance testing, switching systems, and videotex. Short notes, conference reports and summaries. Calendar of international conferences, symposia, and other meetings related to computers, communications, and publishing. Occasional special issues.

1049 **ELECTRONICS AND COMMUNICATIONS IN JAPAN: PART 1, COMMUNICA-TIONS.** New York: Scripta Technica, 1985–date, monthly. English-language translations of highly technical research articles originally published in Japanese in DENSHI JOHO TSUSHIN GAKKAI RONBUSHI on "communications networks, systems, systems, and services, switching and communications processing, signalling systems and communications protocols, communication theory and software, source encoding, transmission equipment and terminals, optical, cable, satellite and waveguide communications, devices and circuits, measurements and metrology, as well as on radio and satellite communicationsm antennas and propagation, and electromagnetic shielding and capability." Translations published 24 weeks after Japanese originals. Part 1 of three-part TRANSACTIONS OF THE INSTITUTE OF ELECTRONICS, INFORMATION AND COMMUNICATIONS ENGINEERS OF JAPAN, Japan's "foremost electronics periodical," which also includes part 2, ELECTRONICS and Part 3, FUNDAMENTAL ELECTRONIC SCIENCES.

1050 **IEEE COMMUNICATION MAGAZINE.** New York: Institute of Electrical and Electronics Engineers, 1953–date, monthly. Formerly IEEE COMMUNICATIONS SOCIETY MAGAZINE (1953–1978). Aimed at electronic communications professionals as both technicians and consumers. Features on ATM, standards, ISDN, "teleaction services" (banking, shopping, home monitoring), video on demand, telephone IC-Cards, LANs, telefacsimile, cellular radio, fiber optics, and research trends in U.S. and elsewhere. Occasional historical features on early telegraphy and telecommunication. Special theme issues ("Lightwave Components," "Neural Networks"). Book reviews, article abstracts ("Scanning the Literature"), "New Products" profiles. Conference calendar and conference announcements, calls for papers, reports on meetings, society news. Extensive job listings in academics and industry.

1051 **IEEE JOURNAL ON SELECTED AREAS IN COMMUNICATIONS.** New York: Institute of Electrical and Electronics Engineers, 1983–date, bimonthly. Analog and digital signal processing, audio and video encoding, design of transmitters, repeaters, computer communications systems, and software.

1052 **IEEE NETWORK.** New York: Institute of Electrical and Electronics Engineers, 1987–date, quarterly. Practical technical articles on computer communications, standards, and development: optical, digital, and broadband computer networks—LANS, WANS, VANS, ISDN, etc.

1053 **IEEE SPECTRUM.** New York: Institute of Electrical and Electronics Engineers, 1964–date, monthly. Formerly ELECTRICAL ENGINEERING (1931–1963). Flagship publication of the IEEE, representing all of the Society's interests. Articles focus on innovations in electronics, with brief features on corporate and government research, Society news, job listings.

1054 **IEEE TRANSACTIONS ON COMMUNICATIONS.** New York: Institute of Electrical and Electronics Engineers, 1972–date, monthly. Formerly IEEE TRANSACTIONS ON COMMUNI-CATIONS AND TECHNOLOGY (1964–1971; and under various other titles from 1953). Research articles on all fields of telecommunications, including modulation and signal design, coding and communication theory, speech, image, video, and signal processing, network management and operations, wireless communication, network performance and teletraffic analysis, transmission systems, optical communication, systems and services, network control and protocols, and network architechture.

1055 **IEEE TRANSACTIONS ON SIGNAL PROCESSING.** New York: Institute of Electrical and Electronics Engineers, 1991–date, monthly. Formerly IEEE TRANSACTIONS ON

ACOUSTICS, SPEECH, AND SIGNAL PROCESSING (1974–1990); and IEEE TRANSACTIONS ON AUDIO AND ELECTROACOUSTICS (1951–1973). Technology and systems for transmitting speech and other signals by digital, acoustic, or optic means.

1056   IEICE TRANSACTIONS ON COMMUNICATIONS. Tokyo: Institute of Electronics, Information, and Communication Engineers, 1991–date, monthly. Formerly IEICE TRANSACTIONS ON COMMUNICATIONS, ELECTRONICS, INFORMATION, AND SYSTEMS (1991). Articles in English on communication engineering technologies, including communications networks, switching and processing, signalling systems and protocols, optical, cable, radio, and satellite communications, terminals and equipment.

1057   INFORMATION & MANAGEMENT: THE INTERNATIONAL JOURNAL OF INFORMATION SYSTEMS APPLICATIONS: SYSTEMS, OBJECTIVES, SOLUTIONS, (SOS). Amsterdam, The Netherlands: Elsevier Science Publishers, 1978–date, 10/year in 2 vols. Publishes research at the interface of organizational communication and new communication technologies. Aimed at managers, professionals, administrators, and executives looking for competitive advantages through enhanced access and control of information via new technologies. Articles on influence of information technology on organizational structure, computer-aided design and user performance, interactive information processing, local area networks, data security, effective telecommunications, and administration of information systems. Conference calendar, announcements, and calls for papers.

1058   JASIS: JOURNAL OF AMERICAN SOCIETY FOR INFORMATION SCIENCE. New York: John Wiley & Sons, 1950–date, 8/year. Official journal of the American Society for Information Science. Research and discussions of information and communication theory and related economic, business, legal, and social implications. Cover articles focus on artificial intelligence, man-machine interaction, economics of information, the design of information technology, the ethics of information, and the impact of information systems and technology on society. Historical features, brief reports and notes on research in Europe, opinion papers, book reviews.

1059   JOURNAL OF COMMUNICATIONS TECHNOLOGY AND ELECTRONICS. New York: Scripta Technica, 1993–date, 16/year. Formerly SOVIET JOURNAL OF COMMUNICATIONS TECHNOLOGY AND ELECTRONICS (1985–1992) and RADIO ENGINEERING AND ELECTRONIC PHYSICS (1961–1984). Highly technical research articles on "theory and physical fundamentals of communications and electronics engineering": "circuit theory and applications, applied mathematics, communication theory, electrodynamics, wave propagation, antennas and waveguides, space communications, signal processing, solid state theory and devises, quantum electronics and electron-optical devices, and electromagnetic theory."

1060   NETWORKS: AN INTERNATIONAL JOURNAL. New York: John Wiley and Sons, 1970–date, 7/year. Highly technical and theoretical articles on "the design and analysis of any large-scale network," including data, video, vehicle, computer or power networks. Features on satellite network design, broadcast protocals, telecommunication engineering. Book reviews.

1061   SPACE COMMUNICATIONS. Amsterdam: IOS Press, 1989–date, bimonthly. Formerly SPACE COMMUNICATIONS AND BROADCASTING (1983–1989). Covers all aspects of satellite communications, including satellite broadcasting, user requirements of satellite systems, satellite and earth station design, specific satellites and systems. Special issues and reports. Schematics and other illustrations.

1062   TELECOMMUNICATIONS AND RADIO ENGINEERING. New York: Scripta Technica, 1964–date, monthly. Combined English-language translation of two highly technical research journals of Russian National Society of Electronics Engineers, ELEKTROSVYAZ and RADIOTEKHNIKA. Particular emphasis on "long-range and satellite communications": "digital and analog wire, radio, video, and optical communications, facsimile, micro- and millimeter-wave communications, switching and coding theory, signal processing, voice and pattern recognition, filters and noise immunity, antennas, waveguides, electron devices and industrial electronics."

8-C-2. House Technical Journals

1063　**AT&T TECHNICAL JOURNAL.** New York: AT&T Bell Laboratories, 1985–date, bimonthly. Formerly AT&T BELL LABORATORIES TECHNICAL JOURNAL (1984) and BELL SYSTEM TECHNICAL JOURNAL (1922–1983). Research articles by AT&T researchers and technical staff on voice and data telecommunications systems and networks. Intended for other AT&T researchers. Issues typically focus on specific technical, policy, or management topic. Recent issues on wireless technology, product design for global markets, TQM at AT&T, broadband ISDN, video communications. Profiles of new products and services, descriptions of tests and development, international alliances and other activities, "AT&T Innovation Briefs." Very attractively illustrated; color graphics.

1064　**AT&T TECHNOLOGY.** New York: AT&T Bell Laboratories, 1986–date, quarterly. Formerly AT&T BELL LABORATORIES RECORD (1925–1985). Articles report on AT&T technical developments and activities. Intended to "inform" AT&T customers "about AT&T products, systems, services; to explain the technologies underlying them; and to discuss broad trends in AT&T technical developments and their applications." Issues focus on ATM, global information superhighway, high speed data services, etc. Handsome color illustrations, charts, tables.

1065　**BRITISH TELECOMMUNICATIONS ENGINEERING.** London: Institute of British Telecommunications Engineers, 1982–date, quarterly. Formerly POST OFFICE ELECTRICAL ENGINEER'S JOURNAL (1908–1982). Brief features on BT company news as well on industry. Highlights new BT products, systems, and R&D, with emphasis on BT's role in U.K. and European telecommunications.

1066　**COMPONENTS.** Erlangen, Germany: Siemens, 1906–date, bimonthly. Formerly SIEMENS COMPONENTS. Features highlight Siemens' R&D in smart chips, switches, instrumentation, and other electronics equipment.

1067　**COMSAT TECHNICAL REVIEW.** Clarksburg, MD: Communications Satellite Corporation, 1971–date, biennial. Significant emphasis in 1990–1992 issues on INTELSAT VI satellite system. Other technical research articles on ITALSAT, EUTELSAT, and other systems.

1068　**ELECTRICAL COMMUNICATIONS.** Paris: Alcatel Alsthom, 1922–date. Formerly published by ITT. Recent issues focus on specific "key" topics: "Cable for Telecommunications," "SDH/SONET: Business Communications Systems," wireless telecommunications, etc., as well as telecommunications in general. Editorials and technical articles. Emphasis on Alcatel systems, products, and services.

1069　**EXCHANGE.** Piscataway, NJ: Bell Communications Research, 1984–date, five issues per year. Designed for general readers as well as Bellcore personnel, this reviews technological developments and applications of the RBOC-owned research facility.

1070　**FUJITSU SCIENTIFIC AND TECHNICAL JOURNAL.** Kawasaki City, Kanagawa, Japan: Fujitsu, 1965–date, 5/year. Highly technical articles on Fujitsu's research in telecommunications, information systems, and specific advanced information technologies.

1071　**GEC REVIEW.** London: General Electric, 1985–date, 3/year. Describes full range of GEC advanced technology R&D.

1072　**HEWLETT-PACKARD JOURNAL.** Palo Alto, CA: Hewlett-Packard, 1949–date, monthly.

1073　**IBM JOURNAL OF RESEARCH AND DEVELOPMENT.** Armonk, NY: IBM, 1957–date, bimonthly. Issues focus on specific topics in computers, information technologies, and other advanced technologies.

1074　**IBM SYSTEMS JOURNAL.** Armonk, NY: IBM, 1962–date, quarterly. Highly technical articles on IBM systems and software development.

1075   **MITSUBISHI ELECTRIC ADVANCE.** Tokyo: Mitsubishi, 1978–date, monthly. Describes Mitsubishi advanced technologies R&D programs.

1076   **NATIONAL TECHNICAL REPORT.** Osaka: Matsushita, 1955–date, bimonthly. Recent issues have included articles on specific topics like audio and systems engineering.

1077   **NEC RESEARCH AND DEVELOPMENT.** Tokyo: Nippon Electric, 1960–date, quarterly. Recent features on NEC research on private networks, telephone systems, imaging, expert systems, etc.

1078   **RCA REVIEW.** Princeton, NJ: RCA, 1936–date, quarterly.

1079   **TELECOM REPORT INTERNATIONAL.** Berlin and Munich: Siemens, 1978–date, bimonthly. "Intended to document the utility of Siemens products." Reports and company news on implemented telecommunications systems and services worldwide, with significant attention on U.S. market. Recent articles on broadband systems, communications terminals and peripherals, test equipment, computer-supported telecommunications applications, data communications, fiber optics technology and applications, ISDN, mobile communications, network management, networking, private and public communications systems, telecommunications cables, transmission systems, and trends.

8-D. Industry and Economics

8-D-1. Management and Economics Research

1080   **INFORMATION AND DECISION TECHNOLOGIES.** Amsterdam, The Netherlands: Elsevier Science, 1988–date, bimonthly. Research articles cover expert systems, integrated information systems, group decision support systems, software, and other topics.

1081   **INFORMATION ECONOMICS AND POLICY.** Amsterdam, The Netherlands: Elsevier Science Publishers, 1983–date, quarterly. "The official journal of the International Telecommunications Society." Aimed at researchers and professionals ("expert consultants and policy-makers"). Sophisticated, theoretical articles on "information and (tele)communications media" economics and policies. Telecommunication in economic development, copyright, intellectual property and public access through photocopying, the costs of evolving telecommunication technologies, incentives for media regulation are typical topics. Case studies of specific problems. Book reviews, announcements, calls for papers.

1082   **INFORMATION PROCESSING & MANAGEMENT.** Oxford: Pergamon, 1963–date, bimonthly. Sophisticated technical scholarship on the theoretical backgrounds for new communication technologies, including basic and applied research on "experimental and advanced procedures in applications of technologies" and "management of information resources and systems." Topics of interest include artificial intelligence, cognitive science, semantics, economics of information, and information processing and transfer. Contributors largely work in the field of computer systems and information science. Special theme issues ("Modeling Data, Information, and Knowledge"). Book reviews. Conference announcements.

1083   **INTERNATIONAL JOURNAL OF INFORMATION MANAGEMENT.** Oxford: Butterworth-Heinemann, 1986–date, bimonthly. Scholarly research articles on "information systems, organizations, management, decision making, long term planning, information overload, computer and telecommunication technologies, human communication and people in systems and organizations." Recent emphasis on information technologies in national development (France, China, Portugal) and in specific industries and services (pharmaceutical industry, banking, medicine). Letters to editor, calendar of meetings and conferences, book reviews.

1084   **JMIS: JOURNAL OF MANAGEMENT INFORMATION SYSTEMS.** Armonk, NY: M. E. Sharpe, 1984–date, quarterly. Theoretical and practical articles on "all aspects of the structure, development, and utilization of management information systems; pragmatic designs and applications; systems development methodologies and other techniques of software engineering; [and] analyses of informational policy making." Emphasis on MIS in decision-making. Special theme issues ("Decision Support and Knowledge-Based Systems").

1085   **JOURNAL OF REGULATORY ECONOMICS.** Norwell, MA: Kluwer Academic, 1989–date, quarterly. Advanced research articles on economics of regulation, with emphasis on policies. Sponsored by corporations, including U.S. telecommunications companies.

1086   **RAND JOURNAL OF ECONOMICS.** Santa Monica, CA: Rand Corporation, 1984–date, quarterly. Formerly BELL JOURNAL OF ECONOMICS (1970–1984). Scholarly research articles on regulated public utilities industries. Interest in regulation, antitrust, pricing and patent issues of telecommunications and new technologies.

8-D-2. Trade and Industry News

1087   **CELLULAR BUSINESS.** Overland Park, KS: Intertec, 1984–date, monthly. Full-text available online on LEXIS/NEXIS (0728). "The marketing, engineering, and news magazine of the cellular industry." Features aimed at interests of cellular industry professionals. Emphasis on new products and technologies as well as current news relevant to the industry. Features on topics like techniques for concealing cell sites, avoiding lightning damage to towers, etc. Regular departments give brief information on business, legal and regulatory, carrier activities. Concise calendar of programs and conferences.

1088   **COMMUNICATIONS NEWS.** Nokomis, FL: Nelson Publishing, 1964–date, monthly. Absorbed WIRE & RADIO COMMUNICATIONS (1883–1964, various titles). Trade newsletter; useful for current legal, business, and technical developments.

1089   **COMMUNICATIONS WEEK.** Manhasset, NY: CPM Publications, 1984–date, weekly. Full-text available online on DataTimes, LEXIS/NEXIS (0728). "Newspaper for the communications industry." Brief features on current trade and legal news ("Top of the News") relevant to the telecommunication industry in general. Notes on full range of telecommunications topics in regular departments for "Network Applications," "Local Area Networks," "Internetworking," "Network Services," "Mobile Computing," "Network Management," etc. "Closeup" features focus on specific companies and products; "Lineup" gives comparative product reviews. "Company in the News" and "Products in the News" indexes.

1090   **HIGH TECHNOLOGY BUSINESS.** Boston: Infotechnology Publishing, 1981–date, monthly. Full-text available online on LEXIS/NEXIS (0728). Formerly HIGH TECHNOLOGY (1981–1987). Features hype new technologies at business and other professionals, emphasizing competitive advantages. Recent articles on tracking fleet vehicles by satellite, use of interactive television in retail markets, super chips, the fiber optics industry, telephone standardization, cellular radio, and the integration of R&D and marketing. Regular departments on international and national business, technological developments, new publications and products, and Washington DC legislative and policy activities. "Newsletter Digest" selectively abstracts other major industry publications, like COMPUTER DAILY, FIBER OPTICS WEEKLY UPDATE, and TELEVISION DIGEST.

1091   **INFORMATION TODAY.** Medford, NJ: Learned Information Inc., 1984–date, monthly. "The newspaper for users and producers of information services."

1092   **PCIA JOURNAL: THE MAGAZINE FOR THE PERSONAL COMMUNICATIONS INDUSTRY.** Englewood, CO: The Business Word, 1994–date, monthly. Covers various mobile and personal communications with reports and articles. Formerly TELELOCATOR (1976–94).

1093   **RURAL TELECOMMUNICATIONS.** Washington: National Telephone Cooperative Association, 1982–date, bimonthly. Emphases on universal services and service regulations. Regular departments cover technology, association personnel.

1094   **SATELLITE COMMUNICATIONS.** Atlanta: Argus Business, 1975–date, monthly. Technology, policy, and management articles on global satellite developments.

1095   **TELEPHONY.** Chicago: Intertec, 1901–date, weekly. Full-text available online on LEXIS/ NEXIS (0728). Major trade weekly including news items on regulation and technology, articles on management aspects of telephony, policy reveiws, etc.

1096   **VIA SATELLITE.** Potomac, MD: Phillips Business Information, 1986–date, monthly. Covers all aspects of domestic and international satellite development and applications in feature articles and news notes.

8-E. Applications/Impact.

1097   **AVC DEVELOPMENT & DELIVERY: FOR AV, VIDEO, AND COMPUTER GRAPHIC PRESENTATIONS.** Woodbury, NY: PTN Publishing, 1992–date, monthly. Formerly AUDIO-VISUAL COMMUNICATIONS (1961–1991). Articles on audio-visual, computer, video, and multimedia technologies (GIS, MIS, designing broadcasts for computers, etc.) aimed at business and management professionals. Useful for descriptions of new products and technologies, applications.

1098   **BEHAVIOR AND INFORMATION TECHNOLOGY.** Basingstoke, U.K.: Taylor and Francis, 1982–date, quarterly. Research studies focusing on man-machine interaction, including information technology hardware and software design; uses of technology for competitive advantage in education, business, and industry; attitudes toward computerization in specific industries; and computer-assisted evaluation of human performance. Book reviews.

1099   **ECTJ: EDUCATIONAL COMMUNICATION AND TECHNOLOGY JOURNAL: A JOURNAL OF THEORY, RESEARCH, AND DEVELOPMENT.** Washington: Association for Educational Communications and Technology, 1953–date, quarterly. Features research articles on "theory, development, and research related to technological processes in education," with emphasis on innovative applications of communication technologies, such as interactive video, computer-based and computer-assisted instruction, and the use of the telephone in tutoring. Reviews of research, conference papers, and other reports from abroad. Abstracts relevant ERIC documents ("Research Abstracts").

1100   **EDUCATIONAL TECHNOLOGY.** Englewood Cliffs, NJ: Educational Technology Publications, 1960–date, monthly. Publishes news and articles about current technological developments intended to enhance and improve instructional effectiveness, including instructional television, the design of instructional media, expert systems, microcomputers, computer networking, distance learning, and instructional simulation. Special issues on specific themes, such as training teachers to use new technologies and "Hypermedia." Calendar of upcoming conferences and meetings. Brief descriptions of new soft- and hardware. Book reviews ("Educational Technology Professional Literature Reviews"), summaries of ERIC documents.

1101   **IEEE TECHNOLOGY AND SOCIETY MAGAZINE.** New York: Institute of Electrical and Electronics Engineers, 1982–date, quarterly. Formerly TECHNOLOGY AND SOCIETY (1978– 1982). Sponsored by IEEE's Society on Social Implications of Technology. Research articles (occasionally historically-oriented) on social issues of electrotechnology, information technology, and telecommunication, including public policy, economics, and education.

1102   **INFORMATION AGE.** Guildford, U.K.: Butterworth Scientific, 1982–date, quarterly. Formerly INFORMATION PRIVACY (1978–1982). Particularly important for features on transborder data flow and global cooperation as well as electronic data security, computer fraud, and computer viruses. Other articles and regular departments on telecommunication policy news. Conference calendar. Book reviews.

1103   **INFORMATION SERVICES AND USE.** Amsterdam: Elsevier Science Publishers, 1981– date, bimonthly. Articles focus on international developments in information and new communication technologies management and applications in business, government, and society in general, with particular attention to information technology design and innovative applications. Frequently discussed topics include bibliographic information retrieval systems, government information policies, and networking. Special theme issues ("The Database Business: Managing Today, Planning for Tomorrow").

1104   **INFORMATION SOCIETY.** London: Taylor & Francis, 1981–date, quarterly. Features research articles on such issues as transborder data flow, government regulation of the communication industry and control of information, the impact of information and information industry development on society, information and economic development, and technology transfer.

1105   **JOURNAL OF COMMUNICATION.** New York: Oxford University Press, 1951–date, quarterly. One of the most prominent journals in academic communication field. Historically important for early research studies of communication. Emphasis on electronic media policy and contents and effects since 1974. Occasional review essays. International communication conference calendar, announcements, and news. Extensive book reviews.

1106   **MEDIA, CULTURE, SOCIETY.** London: Sage, 1979–date, quarterly. Research on "mass media (television, radio, journalism) within their political, cultural, and historical contexts" and on "issues raised by the convergence of the mass media with systems of cultural production and diffusion based upon telecommunications and computing." Issues on specific themes, assembled by guest editors who provide introductory editorial, covering such topics as Western European satellite television and broadcasting, media imperialism and local economic and cultural development, radio history, and public service broadcasting. Book reviews.

1107   **TELEMATICS AND INFORMATICS: AN INTERNATIONAL JOURNAL.** New York: Pergamon, 1984–date, quarterly. Research on "applied telecommunications and information technology," "information resource management," "socio-economic implications," "international issues in communication," and "information policy and legislation." Features on computer graphics, digital radio networks, speech synthesis, voice recognition, data encryption, satellite television, videotex, ISDNs, earth stations, expert systems, and cable subscriptions. Frequent special issues on specific themes ("Telecom Policies in the Bush Administration," "Local Area Networks," "Telecommunication in the Developing World," and "Telecommunication Programs at U.S. Universities"). Regular department on law and regulation ("Legislative and Policy Focus"). Abstracts of new government publications ("NTIS Section") and U.S. and foreign patents ("Patsearch Section"). Book reviews.

8-F. Domestic Policy

8-F-1. Legal and Policy Research

1108   **CARDOZO ARTS & ENTERTAINMENT LAW JOURNAL.** New York: Yeshiva University, Benjamin N. Cardozo School of Law, 1981–date, semiannual. Scholarly articles on policy and regulation, with occasional notes, essays, and comments. Recent features on FCC, intellectual property rights, cable and telecommunication industry competitiveness, and NAFTA and media ownership. Book reviews; indexes of cited cases and statutes.

1109   **COMMUNICATIONS AND THE LAW: A QUARTERLY REVIEW.** Littleton, CO: Fred B. Rothman, 1979–date, quarterly. Major scholarly journal on current and historical topics in communication related to law and policy. International scope. Articles on cable and newspaper cross ownership, direct broadcasting satellites, pay television, and the like. Special issues on specific themes ("Calling It Piracy: Mass Media and Copyright in the U.S. and Canada"). Book reviews.

1110   **COMMUNICATIONS LAWYER.** Chicago: American Bar Association Press, 1983–date, quarterly. "The journal of media, information, and communications law." Sponsored by ABA's Forum on Communications Law. Intended for practitioners as well as academics and professionals. Focus on U.S. telecommunications policy, law, and regulation. Includes a regular current bibliography of communications law. Book reviews.

1111   **COMPUTER/LAW JOURNAL.** Manhattan Beach, CA: Center for Computer/Law, 1978–date, quarterly. Articles, notes, and special features on the legal aspects of new information technology in society. Particular attention on software and information property rights ("The Impact of Digital Technology on Copyright Law"). Special issues on specific topics, including taxing cable television users, cable franchising, patenting computer software, electronic funds transfer systems, computer crime, and international information law and policy.

1112   **COMPUTER LAWYER.** New York: Law & Business, 1984–date, monthly. Focuses on legislation and litigation related to topics like fair use, patents and intellectual property rights, and uses of information technologies in legal practice. Tracks court descisions.

1113   **FEDERAL COMMUNICATIONS LAW JOURNAL.** Bloomington, IN: Indiana University School of Law, and the Federal Communications Bar Association, 1937–date, 3/year. Full-text available online on LEXIS/NEXIS (0728). Formerly FEDERAL COMMUNICATIONS BAR JOURNAL (Washington: Federal Communications Bar Association, 1937–1976). Sponsored by the Federal Communications Bar Association. Most important journal for U.S. telecommunication law. Scholarly articles, essays, and commentaries on media, telecommunication, and electronics policy and regulation. Recent articles on cable television, computer privacy, deregulation and AT&T, private telecommunication networks, copyright fair use and new communication technologies, remote sensing and privacy, and more. Convenient "Articles Digest" abstracts features in other major communication law journals. Book reviews. Volume indexes, with useful cumulative index for volumes 32–41 (1980–1989) in vol 41.4 (1989).

1114   **GOVERNMENT INFORMATION QUARTERLY: JOURNAL OF RESOURCES, SERVICES, POLICIES, AND PRACTICES.** Greenwich, CT: JAI Press, 1984–date, quarterly. Focuses on issues related to the gathering and disseminating of information by government, both domestic and foreign. Aimed primarily at government officials, policy makers, journalists, and lawyers, as well as librarians and information professionals. Of particular value for research on all aspects of U.S. national information policy, including right to privacy, effects of privatization and access to information, and economic competitiveness. Also useful for studies of information collection and dissemination by specific government agencies, such as EPA, the departments of Commerce and Education, the Defense Technical Information Center, and the Office of Management and Budget. Special "Symposium" theme issues ("Electronic Collection and Dissemination of Federal Government Information" and "National Security Controls on Information and Communication"). Occasional literature reviews. Book reviews (including recent government documents).

1115   **HASTINGS COMMUNICATIONS AND ENTERTAINMENT LAW JOURNAL (COMM/ENT).** San Francisco: University of California, Hastings College of the Law, 1990–date, quarterly. Formerly COMM/ENT: A JOURNAL OF COMMUNICATION AND ENTERTAINMENT LAW and variant titles (1977–1990). Scholarly articles, occasionally historically- oriented, on legal and policy issues in all areas of mass communication, particularly the regulation of communication and telecommunication industries. Recent articles on telecommunication regulation and competition, intellectual property rights and computer bulletin boards, privacy and electronic communications, cable franchising, dial-a-porn, wiretapping, and copyrighting computer programs. Regularly features annotated bibliographies and guides to research resources, including Frank G. Houdek's "Nonbroadcast Video-Programming and Distribution: A Comprehensive Bibliography of Law-related Periodical Articles," 9 (Winter 1987): 307–46; "Video Technology and the Law: A Bibliography of Legal and Law-Related Materials on Cable Television, Subscription-Pay Television, Direct Broadcasting Satellites, Videorecording and Videotext," 5 (Winter 1983): 341–98; and James E. Duggan's "Legal Protection of Computer Programs, 1980–1992: A Bibliography of Law-Related Materials," 15 (Fall 1992): 211–95. Special topical issues ("Piracy and Gray Market Imports," "The Divestiture of American Telephone and Telegraph Company"), and annual computer law symposia. Brief notes and case commentaries; abstracts of recent legal articles. Subject and cited case indexes.

1116 **HIGH TECHNOLOGY LAW JOURNAL.** Berkeley, CA: University of California Press, 1986–date, 2/year. Scholarly research and review articles, commentaries and essays, and case notes on the relationship of technology and law, including space policy, copyrighting software, information property rights, and industrial and organizational innovation. John Pinheiro's "Research Pathfinder: AT&T Divestiture & the Telecommunications Market," 2.1 (1987), 303–355, is useful chronology of events, glossary of terms, summaries of principal federal and state cases and legislation, and extensive annotated bibliography of books, reports, and articles. Regular "Legislative Update" summarizes recent federal and state legislation in high technology. Occasional book and software reviews.

1117 **JOURNAL OF SPACE LAW.** University, MS: Lamar Society of International Law, 1973–date, semiannual. "A journal devoted to the legal problems arising out of human activities in outer space." Issues include 2–3 scholarly research articles on satellite communications and other commercial and governmental uses of space. "Recent Publications" is an important bibliography of books, articles, reports, cases, reprints of documents, etc., on space law. Reports on conferences and meetings, relevant Congressional activities; notes on court cases; news briefs and calendar; book reviews.

1118 **JURIMETRICS JOURNAL OF LAW, SCIENCE & TECHNOLOGY.** Chicago: American Bar Association, 1991–date, quarterly. Formerly MULL: MODERN USES OF LOGIC IN LAW (1959–1966); JURIMETRICS JOURNAL (1966–1991). Sponsored by ABA's Special Committee on Electronic Data Retrieval, Section of Science and Technology, Committee on Law and Technology, and University of Michigan Law School; formerly sponsored by Arizona State University Law School. Scholarly articles on the law in relation to science and high technology, with particular emphasis on computers and new communications technologies. Literature reviews. Book reviews.

1119 **PUBLIC UTILITIES FORTNIGHTLY.** Arlington, VA: Public Utilities Reports, 1929–date, biweekly. Variant title: FORTNIGHTLY: THE NORTH AMERICAN UTILITIES BUSINESS MAGAZINE. A standard industry resource which includes scholarly articles, news reports, and updates on PUC activities nation-wide. Includes but is not limited to telecommunications.

1120 **SPACE POLICY.** Guildford, England: Butterworth Scientific, 1985–date, quarterly. Research on multinational and international politics of uses of space. Articles on remote sensing, meteorological satellite networks, telecommunications and national security, and data protection. Studies of national space programs.

1121 **REGULATION.** Washington: American Enterprise Institute, 1977–date, bimonthly. Subtitled as "AEI journal on government and society." Conservative "think tank" views on deregulation across all fields.

1122 **RUTGERS COMPUTER & TECHNOLOGY LAW JOURNAL.** Newark, NJ: Rutgers Law School, 1981–date, semiannual. Formerly RUTGERS JOURNAL: JOURNAL OF COMPUTERS (1979–1980). Recent scholarly articles on NAFTA and intellectual property rights, software protection.

1123 **TELECOMMUNICATIONS POLICY.** Guildford, England: Butterworth Scientific, 1977–date, quarterly, later 9/year. The most important and comprehensive telecommunication policy journal. Research articles on the economics and politics of international telecommunication, national systems and programs, regulations, and new technologies. Emphasis on telecommunication in economic development. Nation- and organization-specific studies predominate (ITU, U.S., EC, Caribbean, Canada, Pacific Islands, Germany, Brazil, France, New Zealand, Hong Kong, United Kingdom, and Italy). Features ranging from telecottages and mobile satellite services as well as media imperialism. Occasional literature reviews, such as Marcellus S. Snow's "Telecommunication Literature: A Critical Review of the Economic, Technological, and Public Policy Issues," 12 (1988), 153–183. Book reviews.

1124 **YALE JOURNAL ON REGULATION.** New Haven, CT: Yale Law School, 1983–date, 2/year. Scholarly articles on regulation and control of the media, with particular historical interest in deregulation, the break up of AT&T, and the implications for the telecommunication industry.

8-F-2. Law and Policy News

See 1037. **COMMUNICATIONS DAILY.**

1125 **KMB VIDEO JOURNAL.** Block Island, RI: KMB Associates, 1984–date, monthly. Issued on videocassettes. "A video journal on telecommunications policies and practices for leaders in industry, government, and education." Issues focus on specific topics.

See 0859 **NATIONAL REGULATORY RESEARCH INSTITUTE QUARTERLY BULLETIN.**

8-G. International

1126 **COMMUNICATIONS INTERNATIONAL.** London: International Thomson Business Publishing, 1974–date, monthly. Full-text available online on LEXIS/NEXIS (0728). Major focus on global competition and competitiveness. Recent cover features on hot technology topics, including ATM, computer telephony, intelligent payphones, mobile telecommunications. Brief articles on full range of telecommunications topics and interests: industry R&D, national policies, international competition, network economics, new services, market saturation. Regular departments monitor international news, upcoming events, new products.

1127 **AFRICA MEDIA REVIEW.** Nairobi, Kenya: African Council on Communication Education (ACCE), 1986–date, 3/year. "A forum for the study of communication theory, practice and policy in African countries." Scholarly research articles on all aspects of communication, including telecommunications. Emphasis on media policies and national development. Many country studies. Abstracts in English and French. Book reviews.

1128 **BULLETIN OF THE EUROPEAN INSTITUTE FOR THE MEDIA.** Dusseldorf, Germany: European Institute for the Media, 1983–date, quarterly. News features on media developments throughout Europe. "Euromedia News" reports brief items. Statistical data, charts. Book reviews.

1129 **CIRCIT NEWSLETTER.** South Melbourne: Centre for International Research on Communication and Information Technologies, 1989–date, monthly. Issues usually include four news features on full range of telecommunications topics relevant to Australia and Pacific region, including information infrastructures, distance learning, telemedicine, international trade relations in information technologies and telecommunications, media economics, intellectual property rights, competition law and policy, teleworking, etc. "Eye on the World" editorials. Announcements for new CIRCIT and other publications. Events calendar.

1130 **COLUMBIA JOURNAL OF WORLD BUSINESS.** New York: Columbia University, Columbia Business School, 1963–date, quarterly. Particular interest in management and economics in international business, from global perspectives. Occasional special issues on specific themes: vol. 22.3 (Fall 1987) focused on "International Commercial Television." Book reviews.

1131 **COMMUNICATIONS AND STRATEGIES.** Montpellier, France: IDATE, 1991–date, 3/year. Major telecommunication research journal featuring authoritative articles either in French or English by major scholars. Emphasis on European telecommunications infrastructures, regulation, competition, economics, and socio-cultural impacts. Numerous country and operator/company studies (British Telecom, Deutsche Bundespost Telecom). Recent emphasis on strategic alliances/joint ventures in telecommunications industry as well as universal service. Book reviews.

1132 **DIGITAL AUDIO BROADCASTING NEWSLETTER.** Geneva, Switzerland: European Broadcasting Union, 1993–date, quarterly. "International News and Strategic Analysis on Digital Sound Broadcasting." National, international (non-European), and industry information. Data on field trials. "For Reference" offers bibliographies of publication related to DAB and EBU. Conference calendar.

1133 **DIFFUSION**. Geneva, Switzerland: European Broadcasting Union, 1958–date, quarterly. Formerly EBU REVIEW (1958–1991). Useful as indicator of EBU's position on convergence of broadcasting media and telecommunication and information technologies: recent features on pan-European satellite channels and computer-aided radio technologies, for example. Many country studies. Book reviews. Glossy illustrations.

1134 **ESPACE**. Geneva, Switzerland: European Broadcasting Union, 1991–date, monthly. News features on EBU and its programs; interviews with media/telecommunications executives and administrators.

1135 **EUROPEAN JOURNAL OF COMMUNICATION**. London: Sage, 1986–date, quarterly. Scholarly articles, often historically oriented, on comparative national and international electronic communication, particularly country-specific policy issues, such as satellite television in Hungary, German cable industry, and Polish broadcasting. Occasional review articles on new communication technologies research. Special issues on specific themes. Book reviews.

1136 **EUROPEAN JOURNAL OF POLITICAL RESEARCH**. Dordrecht, The Netherlands: Kluwer Academic Publishers, 1973–date, bimonthly. "Official journal of the European Consortium for Political Research." Frequently features scholarly, usually comparative, articles on international information policy and telecommunication and electronics law and regulation. Occasional special issues on telecommunication policy ("Deregulation in Western Europe"). Book reviews.

1137 **GLOBAL TELECOMS BUSINESS**. London: Euromoney Publications, 1994–date, bimonthly. Features articles on markets, finance, regulation and special topics—such as "Asia Survey" in the second issue. Aimed primarily at management and the investment community. Excellent graphics, often in color.

1138 **IFIP NEWSLETTER**. Geneva, Switzerland: IFIP Secretariat, ca. 1971–date, quarterly. News features on uses of information technologies and telematics. Conference reports, calendar of IFIP meetings, publications announcements.

1139 **INNOVATION AND TECHNOLOGY TRANSFER**. Luxembourg: CEC, 1988–date, irregular. Glossy, heavily illustrated newsletter promoting DG XIII telecommunication and information technology research pilot programs (VALUE, RACE, SPRINT, etc.). Reports on conferences, announcements. With XIII MAGAZINE (1153), useful for monitoring EC telecommunication initiatives.

1140 **INTERMEDIA.** London: International Institute of Communications, 1973–date, bimonthly. Important for think pieces and research articles by authorities on timely international telecommunications topics: recent features on distance education, telecommunications and development, and electronic superhighways. Country studies. Statistical data. Glossy illustrations. Letters to the editor, book reviews.

1141 **IPDC NEWSLETTER: INTERNATIONAL PROGRAMME FOR THE DEVEL-OPMENT OF COMMUNICATION**. Paris: Unesco/IPDC, 1982–date, irregular. "Produced with the assistance of the Communication Division of Unesco's Communication, Information, and Informatics Sector," news features emphasize information technology development in developing nations. Occasional notes on new publications.

1142 **JOURNAL OF DEVELOPMENT COMMUNICATION**. Selangor, Malaysia: Asian Institute for Development Communication, 1990–date, semiannual. Research articles, notes, case studies, and editorials emphasize use of telecommunications, mass media, and information technologies in North-South development, particularly in rural communities and agriculture. Many studies of telecommunications on different countries. Book reviews.

1143 **KEIO COMMUNICATION REVIEW**. Tokyo, Japan: Institute for Communication Research, 1980–date, annual. Scholarly articles on telecommunication, mass media, and information technologies. Includes many country studies from Japanese research perspectives.

1144   **MEDIA ASIA: AN ASIAN MASS COMMUNICATION QUARTERLY.** Singapore: Asian Mass Communication Research and Information Centre, 1974–date, quarterly. Authoritative research and commentaries on all areas and aspects of mass communication in Asian nations, with significant emphasis on roles of mass communication in economic, political, and social development. Nation specific studies. Also regularly features commentaries on such topics as government regulation of telecommunication, media and industry responsibility, and North-South issues. Several bibliographical departments: short notices of new publications held by the Centre ("Amicinfo") and government publications ("Documentation List"). Book reviews.

1145   **MEDIA DEVELOPMENT.** London: World Association for Christian Communication, 1980–date, monthly. Articles cover the full range of media, including telecommunications. Special issues focus on specific topics: vol. 2/1992, "New Communications for a New Century." Reports on seminars and conferences, occasional special reports, book reviews.

1146   **MEDIA INFORMATION AUSTRALIA.** North Ryde, N.S.W., Australia: Australian Film and Television School, 1976–date, quarterly. Combines journalistic and scholarly articles on all aspects of media and telecommunication with occasional single-topic special issues: no. 71 (February 1994), "Global Media Games," featuring 16 articles on the internationalization of communication. Useful regular departments include contents reviews of recent issues of communication journals (topically arranged); "Media Briefs" digests news from Australian media under variety of topics, including "Media Ownership" and "New Technology." Reports on research in progress, book reviews.

1147   **NEWSLETTER OF THE INTERNATIONAL TELECOMMUNICATION UNION.** Geneva: ITU, 1994–date, 10/year. Replaced TELECOMMUNICATIONS JOURNAL (1152) as main source of current information on the ITU. Brief news features on activities of ITU and its committees, study groups, projects, as well as on international telecommunications development in general. Listing of satellite launchings, conference calendar, notices of ITU personnel changes, job vacancies.

1148   **NORDICOM REVIEW OF NORDIC RESEARCH ON MEDIA AND COMMUNICATION.** Goteborg, Sweden: University of Goteborg and Nordicom, 1994–date, semiannual. Formerly NORDICOM REVIEW OF NORDIC MASS COMMUNICATION RESEARCH (1981–1993). Research articles, brief notes, and literature reviews on the full range of media topics, including telecommunications. Book reviews.

1149   **OCEAN VOICE: MARITIME INFORMATION TECHNOLOGY & ELECTRONICS.** London: INMARSAT, 1980–date, quarterly. Official journal of the International Maritime Satellite Organization features several articles per issue along with many news briefs.

1150   **PROMETHEUS.** St. Lucia, Australia: Information Research Unit, Department of Economics, University of Queensland, 1983–date, semiannual. "The journal of issues in technological change, innovation, information economics, communication and science policy." Scholarly research articles and short notes on global telecommunications, with particular emphasis on Australia and Pacific Rim. Recent features on telecommunications in medicine and agriculture, telecommunications development in specific regions, telework, robotics, organizational structure and high technology innovations. Substantial book review section. Conference and society news.

1151   **TDR: TRANSNATIONAL DATA AND COMMUNICATIONS REPORT.** Washington: Transnational Data Reporting Service, 1978– date, irregular (monthly). Formerly TRANSNATIONAL DATA REPORT (Amsterdam: North-Holland, 1978–1985). Features research on data security and national sovereignty, rights to privacy, deregulation and government control of information, government uses of information, and international information cooperation. Recent articles on U.S.–Canadian telecommunication relations, telecommunication policies of the U.S., Japan, and the Pacific Basin, and new communication technologies in global banking.

1152   **TELECOMMUNICATION JOURNAL.** Geneva, Switzerland: International Telecommunication Union, 1934–1993, monthly. Formerly JOURNAL TELEGRA-PHIQUE (1869–1933). Separate editions in English, French, and Spanish. Until 1994, the official journal of the ITU, now replaced by NEWSLETTER OF THE INTERNATIONAL TELECOMMUNICATION UNION (1147). Will remain historically important for international policy research. Essential for information on the legal, technological, economic, and sociological implications of telecommunication and new communications technologies in global development. Primary source of information for ITU, CCITT, and CCIR activities and international news. Feature stories on such topics as rural communication satellites in Peru, network management in Australia's national network, communication systems of the International Red Cross and the United Nations, ISDN, remote sensing. Articles did not necessarily express the opinions of the ITU. Several useful bibliographic departments included lists of ITU publications for sale and new films and videos ("ITU Film Library"); book reviews; tables of contents of international telecommunication journals ("Review of Reviews"). Conference calendars for ITU and other telecommunication organizations; and new products announcements.

1153   **XIII MAGAZINE**. Luxembourg: Directorate General for Information Market and Innovation, Commission of the European Communities, 1992–date, quarterly. Formerly IM: INFORMATION MARKET (1985–1991). Heavily illustrated features promoting DG XIII's research programs in telecommunications and information technology, like RACE, EUREKA, STAR, DRIVE, etc.

A. Finding Library Materials: Dewey and Library of Congress Classifications for Telecommunication

Introduction

This guide is intended to aid in library shelf searches on most aspects of telecommunications. Almost any library of significance has a guide or other materials on how to best utilize that particular library—and this appendix is intended to supplement such published material.

While there are many schemes for organizing libraries, this guide covers the two most common systems used in academic and public libraries. Indeed, many libraries use both as they decided some years ago to switch from Dewey Decimal to Library of Congress but lacked funds and manpower to totally re-classify existing collections (this guide got its original inception from the difficulties of using such a library). The two systems detailed here are:

DEWEY DECIMAL SYSTEM (DDC): First compiled by Melvil Dewey as the earliest of "modern" systems of library organization, the Dewey system appeared in 1876 and is now in its 20th edition (1990) or revision. It consists of ten classes (000-099 is for general works, 300s are for social science, 600s for technology, 700s for the arts, and the 900s for history, etc.), each covering an area of knowledge. It is still thought one of the best means of organizing a small-to-medium-sized library because it can expand with the collection, although it lacks the flexibility for a truly large collection. Citations in the Dewey system consist only of numbers, often carried to two or three decimal places as human knowledge has expanded.

LIBRARY OF CONGRESS SYSTEM (LC): As the largest library in the country and one of the biggest in the world, the Library of Congress needed a completely expandable classification scheme and so several people over many years devised the letter-number scheme now in use. It is a far more complicated and detailed system and is constantly under revision to this day. Citations in the LC scheme always begin with one or two letters and then numbers (often up to four digits or more), so they are easily identified.

There are, of course, too many other systems (Cutter, Richardson, etc.) to cover all of them here. Most government documents are filed by a classification scheme established by the issuing agency (state, U.S. government, League of Nations, United Nations, etc.) and are ordered by agency rather than subject. Each library handles documents differently (some file them separately, others intermix them in the stacks using the overall library classification scheme)—you will simply have to check in each case. The same holds true for periodicals—some libraries file them in a single place by title, others separate them throughout the library by subject—again, you have to check. Most libraries maintain reference collections, sort of a cross-section of the full collection filed on a non-circulating basis—and the only copy of a telecommunication reference may be there.

This guide is based on the published detailed descriptions of both Dewey and LC (the latter supplements amounting to over fifty volumes) rather than on books on the shelf because each library takes a certain amount of liberty with the schemes. Misclassification in telecommunications is common, so check any related category you can.

Using This Guide

To be a useful guide, a subject division must be used which by its nature tends to over-compartmentalize an integrated whole. On the following pages will be found sections devoted to each of the following broad subject areas:

1. Bibliographies
2. History
3. Technology
4. Military Communication
5. Organization and Economics
6. Effects and Influence
7. Law and Regulation
8. Foreign and International

The listing is divided into two columns, the classification system for the Dewey system on the left, and LC on the right. Following most specific classifications is a brief parenthetical notation very roughly explaining what will be found there. Remember, the lists are suggestive rather than exhaustive, so check related lists as indicated, depending on the topic you are searching.

1. Bibliographies

Here, particularly, LC is more complicated than Dewey in that the former has a wide spread of subject categories while the latter crowds all bibliographies under one number.

Dewey   **LC**

Telegraph and Telephone

016.384 (general aspects)   Z 5834.T44 (general)
016.621 (technical)   Z 7164. P85 (post, telegraph, telephone)

2. History

Historical references are of two types—technical and general, and will usually be found in at least two places under either classification scheme.

Dewey   **LC**

Telegraph

384.1 (business aspects)   HE 7631 (general)
384.4 (submarine cables)   HE 7711-7741 and TK 5605-5651
                                                      (submarine cable telegraphy)
                                                   HE 7775 (general, in the U.S.)

Telephone

384.6 (business aspects)   HE 8731 (general)
621.385-7 (technical)   TK 6015-6143 (technical)

3. Technology

This category overlaps somewhat with history, but is intended to stress the current state of the art. Check here and in history for full coverage.

Dewey   **LC**

Sound

534 (all aspects)   QC 221-246 (all aspects)

Optics

535 (all aspects)   QC 350-495 (general)
                                                   QC 449 (laser)

Telegraphy

384.15 (facilities)   TK 5105-5599 (general)
621.382-3 (general)   TK 5605-5681 (submarine cable)

Telephony

384.65 (facilities)   TK 6001-6195 (general)
621.385-7 (general)   TK 6201-6525 (facilities, equipment)

Space

621.384 197 (radio in)   TL 796-799 (satellite uses)
621.388 9 (television in)   TL 3025-3040 (communications in space)
629.437, 457 (space communications)

4. Military Communication

Materials here are divided more by period (World War I or II, for example), and by service, rather than by medium. See also foreign and international listings, p. 178.

Dewey   **LC**

Army and Air Force

355.27 (mobilization of   UG 570-620
   communication facilities)   (Army electronic media types and uses)
355.415 (tactical communica-
   tions techniques, etc.)
358.24 (communications
   services)
623.73 (engineering aspects)

Navy

359.98   VG 280-85 (signaling)
                                     VG 70-85 (naval communications)

5. Organization and Economics

Included here are management, finance, industry organization and manufacturing, operations, personnel, and related factors.

Dewey   **LC**

Telegraphy

384.1 (general)   HE 7601-68 (general, administrative)
384.4 (submarine)   HE 7681-95 (rates and finance)
                                       HE 7709-11 (submarine cables)
                                       HE 7761-98 (U.S. companies, etc.)
                                       TK 5283-86. 5381 (technical management)

Telephony

384.6 (general)   HE 8701-35 (general, administrative)
                                       HE 8761-8846 (rates and finance)
                                       TK 6183-86, 6181 (technical management)

6. Effects and Influence

Category includes user information, general influence, relation to other fields, etc.

Dewey   **LC**

Telegraph

384.1 (general)   HE 7631 (general)
                                       HE 7651 (relationship to society, role)

Telephone

384.6 (general)   HE 8731 (general)
                                       HE 7651 (relationship to society, role)

7. Law and Regulation

All aspects of law, regulation, and industry responsibility are included here. Many of the classifications listed below will be found in separate law libraries on many campuses.

Dewey	LC
General	
384 (general)	KF 2761-64 (general)
	KF 2765 (FRC and FCC)
	TK 215-255 (legislation)
Telegraph	
384.12-13 (government control)	HE 7645-47, 7761-67, and 7781-83
	(policy and laws)
	KF 2775 (general and specific)
Telephone	
384.62-63 (government control)	HE 8741-49, 8778 (policy)
	KF 2780 (general and specific)
Radio	
384.522-3 (radio-telegraphy)	HE 8666-67 (wireless)
Space communications	
341.52 (space law)	JX 5810 (diplomatic aspects)

8. Foreign and International

References here are for foreign (domestic communications within a single country), and international (across borders of more than one country) communications.

Dewey	LC
Telegraph	
384.1 (by country)	HE 7700-7705 (world agencies, law, etc.)
621.382 8 (cable systems)	HE 7709-7741 (submarine cables)
	TK 5605-5681 (submarine cables, technical)
	HE 7811-8630 (by country)
	TK 5121-24 (by country, technical)
Telephone	
384.6 (by country)	HE 8748-49 (legal aspects)
621.385-7 (technical)	HE 8861-9680 (by country)
	TK 6021-24 (specific countries, technical)
	TK 6157 (training and study abroad)
Radio	
384.5 (by country)	HE 8665 (wireless in specific countries)
621.384 151 (short-wave)	HE 8668-70 (international law, treaties)
	TK 5720-28 (telegraphy, by country)
	HE 9711-15 (wireless telephony)
	TK 6548 (telephony, by country)

B. Selected Library of Congress Subject Headings for Telecommunication

Based on LIBRARY OF CONGRESS SUBJECT HEADINGS (Washington: Library of Congress, Cataloging Distribution Service, 1994), 4 vols., containing "subject headings created by catalogers and used at the Library of Congress since 1898" (p. iii). Users should also consult the "Introduction" (pp. vii–xvii) for additional guidance regarding subject headings for personal names, corporate bodies, jurisdictions, and other kinds of headings.

Advertising—Telecommunication industry
Advertising—Telephone companies
Advertising—Telephone supplies
Aeronautics—Communication systems
Aerospace telemetry
Artificial satellites in telecommunication
Audiotex
Automated tellers
Biotelemetry
Broadband communication systems
Cables, Submarine
Caller ID telephone service
Cellular radio
Cipher and telegraph codes
Coaxial cables
Coherer
Collective bargaining—Telecommunication
Collective bargaining—Telegraphers
Collective labor agreements—Telecommunication
Command and control systems
Communication and traffic
Computer-assisted instruction
Computer conferencing
Computer networks
Crosstalk
Data transmission systems
Decision support systems
Decremeter
Dial-a-message telephone calls
Digital communications

Distance education
Eccard telephone
Electric power systems—Communication systems
Electronic funds transfer
Electronic mail systems
Electronic measurements
Emergency communication systems
Emergency medical services—Communication systems
Facsimile transmission
Freedom of information
Government communications systems
Home banking services
Hotel communication equipment industry
Hotlines (Counseling)
Image transmission
Information storage and retrieval systems—Telecommunication
Information technology
Interactive television
Intercommunication systems
Interstellar communication
Laser communication systems
Laser communication systems industry
Medical electronics
Medicine—Communication systems
Microwave communication systems
Microwave transmission lines
Military telecommunication
Military telegraph
Military telephone
Millimeter wave communication systems
Mine communication systems
Mobile communication systems
Morse code
Multichannel communication
Multiplexing
Optical communications
Packet switching (Data transmission)
Phase-locked loops
Phonic wheel
Phototelegraphy
Police communication systems
Private branch exchange telephone equipment industry
Pulse modulation (Electronics)
Pulse techniques (Electronics)
Radio
Radio frequency allocation
Radio in education
Radio lines
Radio paging equipment industry
Radio telemetry
Radioteletype

Radiotelephone
Random Access Measurement Systems
Rural telecommunication
Rural telephone
Scrambling systems—Telecommunication
Selling—Telephones
Sequence theory
Signal theory (Telecommunication)
Speech processing systems
Speech scramblers
Spirit phone calls
Spread spectrum communications
Statistical communication theory
Strikes and lockouts—Telegraph
Strikes and lockouts—Telephone companies
Strip transmission lines
Switching theory
Talking yellow pages
Tariff on telecommunications equipment
Telecommunication
Telecommunication cables
Telecommunication equipment industry
Telecommunication equipment leases
Telecommunication in art
Telecommunication in education
Telecommunication in higher education
Telecommunication in libraries
Telecommunication in medicine
Telecommunication lines
Telecommunication policy
Telecommunication switching systems equipment industry
Telecommunication systems
Telecommunication systems testing equipment industry
Telecommunications libraries
Telecommunications services for the deaf
Telecommuting
Teleconferencing
Teleconferencing equipment industry
Teleconferencing in education
Telegraph
Telegraph, Carrier current
Telegraph, Wireless
Telegraph cables
Telegraph code addresses
Telegraph lines
Telegraph keys
Telegraph stamps
Telegraph stations
Telegraph wire
Telegraphers
Telegraphers' cramp

Telegraphone
Telemarketing
Telematics
Telemeter
Telephone
Telephone, Automatic
Telephone, Carrier current
Telephone, Dial
Telephone, Pushbutton
Telephone, Wireless
Telephone answering and recording apparatus
Telephone answering and recording equipment industry
Telephone answering services
Telephone assistance programs for the poor
Telephone bill paying services
Telephone cables
Telephone communication channels
Telephone companies
Telephone credit cards
Telephone etiquette
Telephone fund raising
Telephone holding companies
Telephone in business
Telephone in church work
Telephone in education
Telephone in job hunting
Telephone in medicine
Telephone in mining
Telephone in politics
Telephone lines
Telephone operators
Telephone reference services (Libraries)
Telephone relays
Telephone repeaters
Telephone selling
Telephone stamps
Telephone stations
Telephone supplies industry
Telephone surveys
Telephone switchboards
Telephone switching systems, Electronic
Telephone systems
Telephone wire
Teleports
Teleshopping
Teleshopping equipment industry
Teletext systems
Teletype
Teletype in aeronautics
Television
Television frequency allocation

Toll-free telephone calls
Trade unions—Telecommunication
Trade unions—Telegraphers
Trade unions—Telephone company employees
Underground telephone lines
Used telecommunication equipment
Video conferencing
Video telephone
Videotex systems
Voice mail systems
Wages—Telecommunication
Wireless communication systems
Work groups—Data processing

Index of Main Entries

(Note: Titles are only given for those works lacking specified authors.)